NASA
SPACE SHUTTLE

1981 onwards (all models)

In memory of Aaron Cohen (1931–2010), Shuttle Orbiter Manager 1972–1982, Director of the NASA Johnson Space Center (1986–1993)

First published in April 2011
Reprinted in December 2016, April and November 2018, April 2019, February 2021, December 2022 and August 2023

A catalogue record for this book is available from the British Library

ISBN 978 1 84425 866 6

Library of Congress control no. 2010934899

Published by Haynes Group Limited, Sparkford, Yeovil, Somerset, BA22 7JJ, UK.
Tel: 01963 440635
Int. tel: +44 1963 440635
Website: www.haynes.com

Haynes North America Inc.,
2801 Townsgate Road, Suite 340, Thousand Oaks, CA 91361, USA.

Printed in Malaysia.

COVER CUTAWAY:
NASA Space Shuttle. *(Courtesy of Flightglobal Archive – www.flightglobal.com)*

NASA SPACE SHUTTLE

1981 onwards (all models)

Owners' Workshop Manual

An insight into the design, construction and operation of the NASA Space Shuttle

David Baker

Contents

OPPOSITE The world's only winged spacecraft able to carry astronauts. The legacy of the Shuttle is out there in the form of the Hubble Space Telescope, the International Space Station and a host of satellites and spacecraft sent on their way by this unique vehicle. *(NASA)*

Introduction

At precisely 7.00am local time on 12 April 1981, the first machine to unite the two engineering concepts of aircraft and spacecraft into a single entity lifted off. It was called the Space Shuttle. Little more than 77 years earlier a powered flying machine had carried a human occupant through the air for the first time, and it was 20 years to the day since the first man had been sent into space.

I had first encountered the Shuttle in 1974 at Palmdale, California, when the manufacturer (then known as North American Rockwell) cut metal and began assembling the first Orbiter. But I had already lived with it for eight years as an idea, an evolving concept from the days before moon flights, brought to fruition while humans were still driving electric cars around the lunar surface.

The Shuttle had been a long time coming and it had gone through a period of metamorphosis, transformed from the configuration favoured by its advocates as far back as the 1950s. Compromised by politics, driven forward by ambitious goals but dramatically under-funded, the Shuttle emerged as a compromise between competing requirements, needs and specifications from the customers that planned to use it. Yet there it was, gleaming white and awesome in its power and potential as the world's first reusable space transportation system. And it was as impressive as it was influential in garnering popular support from a wider community than those traditionally associated with space flight. Even the sci-fi world got involved, playing a pivotal role in naming the first Orbiter. When NASA planned to call OV-101 *Constitution*, a public campaign began that included most of the cast from television series *Star Trek,* which saw the name changed to *Enterprise* after the fictional starship!

That the Shuttle was late in flying was a surprise only to those unschooled in what it takes to develop and test a new aerospace concept, and when it did fly it trounced the doomsayers and proved itself a magnificent flying machine. It took guts to fly *Columbia* on its first mission into space. Nothing like it had been built before and when John Young and Bob Crippen piloted the Shuttle into space for the first time they were riding the product of calculation and extrapolation beyond what had ever been tested. Since then two Shuttle Orbiters have been destroyed in avoidable disasters, and it is said by critics to have failed the expectations of those who designed and built it. If it has failed as a vehicle it is because it has been let down by the people who managed it, and not the astronauts who flew it. If it has succeeded it is sometimes in spite of flawed management and ill-advised decisions in an agency that has changed more fundamentally than those on the outside can imagine.

When the Shuttle was conceived, NASA was a government agency run by a young cadre of engineers, technicians and scientists whose average age was around 25. It was a legacy of the National Advisory Committee for Aeronautics which, since the end of the First World War, helped write the rules on aircraft design and engineering. Over the 40 years

BELOW *Discovery* **touches down at Edwards Air Force Base, a symbol of everything the Shuttle has been for more than 30 years – the world's only reusable spacecraft capable of landing like an aircraft on a conventional runway.** *(NASA)*

spanned by Shuttle manufacture and flight operations, NASA slowly but inexorably shifted to an organisation run by managers, politicos and overseers – and maybe that is the price that had to be paid through those decades of change both inside and outside NASA. But the unassailable truth is that the maturing approach to Shuttle operations has itself been a stabilising influence, helping to restore levels of constraint and safety that right through to the last flight left the Shuttle better equipped for its mission than it had ever been.

NASA launched the Shuttle concept in the late 1960s as a cost-saving means of routine transportation to low earth orbit. From there, fleets of moon-ships would depart for bases in lunar orbit and research facilities on the surface of our nearest celestial neighbour. Other vessels, assembled from sections lifted by the Shuttle, would leave earth orbit for the planet Mars, or reach deeper into the solar system to visit the asteroids and send space probes to the outer planets where conditions are too hostile for a human presence. Astronauts would step down onto the red soil of Mars, while teams of up to 100 scientists would fly to their workplace in giant space stations, or make long journeys across the lunar surface, mapping and sampling the rocks as they went.

Within three months of landing on the moon for the first time, NASA delivered these objectives for what it saw as its goal for the next two decades, hoping to achieve all these things and more by the end of the 1980s. In the end none of that could be delivered and even the assembly of the space station, which the Shuttle was built to support, only began in earth orbit 17 years after the first Shuttle flight. Essentially complete in 2011, the International Space Station is very different from the one envisaged by NASA in 1969. Instead of an icon of American supremacy in space technology, it is now the product of the world's greatest international engineering enterprise, bringing together engineers and scientists from a host of countries across Europe as well as Russia, Japan and Canada – in addition to the United States. Former adversaries in a Cold War have warmed to the notion of international cooperation and that has been the hallmark of the Shuttle, for some achieving its true value outside the domain of space science and engineering.

It was a bold day indeed when President Bill Clinton extended a hand of cooperation to the main architect of the former Soviet empire, bringing together the greatest and most illustrious team of space engineers from a very different era. They were now working together, not apart, and forging a new way of building trust and a lasting bond between scientists and engineers across what so recently had

ABOVE On pillars of fire, the Shuttle is launched towards space, bridging earth with the cosmos and carrying the dreams of many who worked tirelessly for decades to develop and operate the most awesome flying machine ever built. *(NASA)*

private industry to come up with crew vehicles for carrying people to and from the International Space Station, now expected to continue in operation for at least another decade. Privately owned and operated fleets of cargo freighters will ply between launch sites and the station to keep it replenished, while NASA works on a vehicle that could carry humans back to the moon and on to Mars – even to the asteroids.

But the Shuttle is a very hard act to follow, not simply because it has been around for such a long time. It has entrenched a way of doing things that will be hard to change. One aspect remains: the determination to succeed and to learn from past efforts. As expressed by President John F. Kennedy in a speech at Rice University, Texas, when speaking of Apollo in 1962, 'we choose [to do these things] not because they are easy but because they are hard…' Nothing about the Shuttle has been easy and space flight has been hard to do. It is inherently dangerous and it is not routine, and probably never will be in the lifetime of anyone reading these words. People have been killed and we will lose more lives in making spaceflight safer, but the Shuttle has provided a valuable tool in advanced aerospace engineering and has pushed the boundaries, in some ways beyond those defined when the programme began. And as already stated, it has been around for a long time.

It is a measure of the Shuttle's longevity that in the last few years of its flight operations very few technicians working to prepare it for launch at the Kennedy Space Center knew much about its origin, or the sweat and tears offered up when it was designed this way more than 40 years ago. In fact, very few of the young engineers who now work on it were even born when it first flew. Yet anyone who has stood alongside, let alone sat inside, this living, breathing entity, hissing, popping, seeming to breathe life from the very depths of its interior, can have failed to have been moved by the sheer scale of this remarkable aerospace achievement. In every essence, to the very last, an incredibly beautiful flying machine.

David Baker
Sussex, England
January 2011

been an unbridgeable gulf. The success of the Shuttle is beyond lost dreams and failed hopes of the generation who brought it to fruition, and it has outlived the bi-polar world of political challenges that funded its development. The Shuttle is the legacy of all that went before, yet it bequeaths a wealth of practical knowledge and understanding about rocketry, the science of flight at high speed through the outer atmosphere, and about routine flight operations with a reusable vehicle.

It is unlikely that the Shuttle will have a successor as big or as ambitious in its goals as the Shuttle was, in comparison to what had gone before. But its contribution to aerospace engineering is immense. Many of the lessons learned will not be forgotten and the next generation of air and space vehicles will be better because of what the Shuttle has pioneered – details of manufacture, assembly and operation that could never have been learned in any other way. Its successor will not be a single vehicle but a range of competitive commercial spacecraft, built by a new generation of aerospace engineers.

In 2010 President Barack Obama stripped NASA of primary responsibility for building the new generation of spacecraft. Instead he tasked the space agency with stimulating

Acknowledgements

In no order of precedence or priority, I would like to thank a few among many who over the last 45 years have played significant roles in helping me through various stages of my career. Specific to the Shuttle programme, however, I would like to acknowledge the seminal guidance of Max Faget in helping me appreciate the nuances of aeronautical design and engineering; and of Aaron Cohen, Manager of the Shuttle Orbiter office (1972–82), whose work was an inspiration and whose influence on my own career was legion. Both Max and Aaron were giants in an age of great aerospace designers, who helped raise the bar on aeronautical and space achievements in engineering and design philosophy.

In the engineering and manufacturing world, many have helped with details on assembly and test including Bob Biggs, Paul Castenholz, and many others too numerous to mention at (the then) North American Aviation and at Rocketdyne, at General Dynamics and at Martin Marietta. I would like to thank the NASA Chief Engineer during the 1980s, Stan Weiss, who worked tirelessly in accommodating my requests for him to brief visitors to NASA headquarters from both Europe and from the USA, anxious to invest in the Shuttle era and its promise of commercial launches. I would like to thank Sy Rubinstein at Rockwell and the late Capt Chester M. 'Chet' Lee at NASA for helping me serve as intermediary between potential satellite customers and the Office of Manned Space Flight at NASA HQ, and for working with me on space commercialisation back in the days when we saw the Shuttle as vital to that effort.

I must also acknowledge the tireless enthusiasm of NASA Administrator James M. Beggs and the many very early morning meetings to review the international opportunities for the Shuttle, and to Gen James A. Abrahamson, with whom potential improvements to Shuttle systems were discussed at length. For encouragement to keep writing, I would like to thank David R. Scott, veteran of three space flights including command of the Apollo 15 moon mission. For opening doors into the world of intelligence gathering from space, thanks are due to Gen George J. Keegan Jr, head of US Air Force Intelligence with whom, during the 1970s, I was involved in analysing Soviet activities that ultimately led to my involvement in the military potential of the Shuttle. To him, thanks for my involvement in Operation Igloo White.

Over the last several decades of my professional career I have been aided and encouraged by various members of the public relations departments of the companies I have been involved with, and they have provided much of the visual material contained in this book. In particular I would like to thank the encouragement of the late William J. O'Donnell at the Kennedy Space Center; the PAO team at NASA's Johnson Space Center; Joseph M. Jones (now retired) formerly from the Marshall Space Flight Center; Sue Cometa at Rockwell; the late Joyce Lincoln, formerly of Rocketdyne; and many others who know that I remember and thank them.

I would like to make special mention of the historians at NASA – of the considerable assistance over time given by Dr Roger Launius, formerly Chief NASA Historian and now at the Smithsonian Air and Space Museum; and of the current staff at NASA HQ, as well as the history team at the Kennedy Space Center. To you all, my sincere thanks and appreciation for great help over many years.

Finally, but most certainly not least, I would like to thank my long-suffering publisher Mark Hughes together with Jonathan Falconer and Steve Rendle, along with all the crew at Haynes, who worked with great patience to bring this book to fruition. There are no mistakes in the book other than those I alone am responsible for.

Chapter One

Genesis

The Shuttle project was born before astronauts reached the moon, and it grew into a potential replacement for all conventional rockets. The idea for a reusable space transportation system was conceived less than a decade after the first satellite, *Sputnik 1,* was launched in 1957. But it would take another decade to bring the Shuttle to maturity and longer still to get it into space. Ahead lay a tortuous path, as politicians, budget officials and lapsed customers took charge of its development, creating something very different to NASA's original vision.

LEFT Facilities built in the 1960s for Apollo moon landings were modified for the Shuttle during the late 1970s. Here a full-scale mock-up of a Saturn V is moved to the launch pad on a Mobile Launch Platform by the Crawler Transporter, equipment that would be used for all Shuttle flights. This was the first full-scale test for all the Apollo ground handling and pad facilities on **25 May, 1966.** *(NASA)*

ABOVE The Saturn V had a height of 363ft when fully fuelled. With almost the same amount of thrust, the Shuttle would be only half that height and just two-thirds of the weight. *(NASA)*

When Neil Armstrong became the first human to set foot on the moon on 21 July 1969, many people thought it marked the start of a bold new adventure, where spaceships carrying brave astronauts would boldly go where nobody had ever gone before. It had been little more than 12 years since the first satellite – Russia's *Sputnik 1* – had bleeped its way across the sky, its monotone a starting pistol for what would be one of the greatest challenges

LEFT Less than eleven years after *Sputnik 1* shocked Americans into a moon race, in July 1969 Neil Armstrong became the first person to stand on the lunar surface, vindicating President Kennedy's boast to beat the Russians. *(The Washington Post)*

of all time: a race to be the first and the best in technology and scientific achievement.

In reality it had little to do with science, but a lot to do with demonstrating technological virility as two very different ideologies vied with each other for the top spot on the global flagpole of political supremacy. The first landing on the moon came a little over eight years after Yuri Gagarin became the first human to orbit the earth, an event that fired the imagination of a generation and became an inspiration for President John F. Kennedy to give NASA the challenge of beating the Russians to the moon.

NASA – the National Aeronautics and Space Administration – had been formed out of an existing US government research institution, the National Advisory Committee for Aeronautics, or NACA. Formed in 1915, NACA became NASA on 1 October 1958 in direct response to *Sputnik* and from then on its challenge was to organise a robust programme of scientific achievements in space science with the goal of making the United States 'pre-eminent' in the exploration of space. NASA inherited plans from

the NACA for a manned spacecraft, a capsule without wings boosted into space by converted ballistic missiles originally developed to lob atomic warheads to the Soviet Union in the event of all-out war.

For many at the NACA it was a disappointment that test pilots would be carried into space in a capsule rather than ascending in a winged space-plane that could be used again many times over. More than a decade of research had gone into the development of rocket powered research aircraft, pushing first beyond the speed of sound and then to the very edge of space itself where air-breathing engines could not operate.

On 14 October 1947 Charles E. 'Chuck' Yeager became the first man to pilot an aircraft through the speed of sound, Mach 1, in a rocket-powered Bell X-1, and by the time *Sputnik* had been launched almost exactly ten years later, experimental rocket propelled research aircraft had exceeded Mach 3. Development was already underway on the North American X-15, capable of exceeding

ABOVE Piloted by Maj. Charles E. 'Chuck' Yeager, on 14 October, 1947, the experimental Bell XS-1 became the first aircraft to break the sound barrier in level flight. This aircraft (46-063) was the second of three, later converted to Bell X-1E specification, carried into the air by a modified B-50. *(NASA)*

ABOVE Neil Armstrong stands at the nose of the North American X-15 in which, as a test pilot, he made seven flights, achieving a maximum speed of more than 5.7 times the speed of sound and a maximum altitude of 39.2 miles. *(NASA)*

RIGHT With a length of 50.75ft and a wing span of 22.3ft, the X-15 was powered by a 57,000lb thrust Thiokol XLR-99 rocket motor – the first capable of being throttled and the first certified for human flight. *(NASA)*

Truly precursors to the Shuttle, three X-15 aircraft made a total of 199 flights between 1959 and 1968, achieving a maximum speed of 4,519mph and a maximum altitude of 67 miles. Plans to send an X-15 into space were cancelled when NASA developed the Mercury spacecraft. *(NASA)*

LEFT Operating on the toxic cocktail of liquid oxygen and anhydrous ammonia, the development of the X-15's Thiokol XLR-99 was a major achievement at a time when rocket motors were notoriously unreliable. *(NASA)*

RIGHT An engineer by training, the Austrian Eugene Sanger was born in 1905 and died in 1964, but in the late 1930s he designed a spaceplane capable of sub-orbital flight around the world. *(David Baker)*

BELOW Dubbed the 'Antipodal Bomber' for its capacity to hit targets on the other side of the world, Sanger's Silbervogel (Silver Bird) was the inspiration for post-World War Two spaceplane concepts. *(David Baker)*

Mach 6 and reaching more than 50 miles above the surface of the earth. So high, in fact, that when the X-15 achieved its goals in the 1960s some of its pilots, in a group that included Neil Armstrong, received astronaut's wings! But it was not only the X-15 research aircraft that was pushing ever faster and higher.

The US Air Force had a project called Dyna-Soar, an unfortunate choice of name but one derived from the principle upon which it was designed to operate, called 'dynamic soaring'. Before the formation of NASA, when the old NACA was solidly employed as the government research agency for aeronautics and high performance flight, the USAF began development of Dyna-Soar as a hypersonic weapon system capable of going into space. In a joint development with the NACA, in 1958 the Air Force began what was expected to be a research vehicle leading to a fully operational space plane.

It had been a long time coming. In the 1930s, the Austrian aerodynamicist Eugene Sanger and his research assistant, later wife, Irene Bredt, had designed such a space plane known as the Antipodal Bomber, deriving that name from its ability to strike targets on the other side of the world from bases in central Europe. Germany funded some development but work slowed when all priority went on the V-2 ballistic missile developed by a team under the charismatic Wernher von Braun. It was left to the victorious Allies to pick up the pieces from war-torn Europe and for the US Air Force to develop a series of proposed developments of the Antipodal Bomber that eventually led to the Dyna-Soar project.

So it was that when *Sputnik 1* was launched and the space age began, the need to compete with the Soviet Union at a time of heightened tensions in the Cold War overthrew the logical progression toward winged space planes. For the immediate future, both sides seemingly picking up momentum in a race to be the best, pace was of the essence. There was no question, man-carrying capsules launched on converted ballistic missiles would get US astronauts into space much earlier than winged space planes, which, with all their sophistication and complexity, would take at least a further decade to develop. So, the first man-carrying space vehicles were capsules and when Russia became the first nation to launch a man into space in April 1961 it was a capsule that would be used to get the first Americans to the moon. Space planes would have to wait.

New shapes for space

The Apollo moon programme dominated the US space budget throughout the 1960s. But as the moon landing goal neared plans were laid for a 'post-Apollo' space progamme, built around future use of Apollo spacecraft, for supporting a permanent scientific base on the moon and using left-over hardware to construct earth and lunar-orbiting space stations. But just as these ambitious plans were being laid down in the middle of the 1960s, the budget began to fall until there was no foreseeable way in which the use of rocket stages thrown away during launch and spacecraft used once only could be paid for.

Kraftstofftanks   *Raketenmotor mit Hilfsmaschinen*

Druckfeste Kabine   *Kraftstofftanks*   *Bombe*   *Eingezogenes Fahrgestell*

RIGHT Deemed not at all fanciful, during the early 1960s the idea of a reusable booster rocket carrying a spaceplane to orbit seemed logical, given the progress in rocketry and manned flight. *(NASA)*

To design a space programme around lower budgets, NASA chose to build its future around reusable systems. Essential to everything was a low-cost, reusable, shuttlecraft carrying astronauts to earth-orbiting space stations from where other reusable vehicles could shuttle people to the moon and back and even begin the exploration of Mars. NASA put together a plan whereby a reusable shuttle and an earth-orbiting space station would be the core around which the future space programme would be built. Everything it did in space after the Apollo moon landings would depend upon an economical transportation system.

It was in London on 10 August 1968 that NASA chose to publicly talk about a reusable shuttle for the first time, when its head of manned space flight, George Mueller, presented a paper to the British Interplanetary Society, the BIS. Formed in 1933, this organisation had been a founding member of the International Academy of Astronautics, the prestigious world body representing space and rocket engineers and scientists. The BIS gave Dr Mueller an award for his work on Apollo and it was then that he chose to articulate what had already been part of NASA planning for at least two years.

Since its inception in 1958, NASA had refused to give up on the space plane concept as a way of developing shuttlecraft for low-cost space transportation. Inheriting from the former NACA aerodynamic studies of hypersonic vehicles capable of flying in excess of Mach 5, the space agency had worked with the US Air Force to develop, build and fly experimental

RIGHT To test how unusual shapes ideal for surviving re-entry from orbit could be piloted through the lower atmosphere to a controlled landing, NASA developed a series of Lifting Body aircraft, this one being the M2-F1. Test pilot 'Chuck' Yeager is at right. *(NASA)*

From left to right, three very different Lifting Body design concepts: the Martin X-24A, the Northrop M2-F3 and the Northrop HL-10. *(NASA)*

BELOW When NASA asked the aerospace industry to work on potential design configurations for a future shuttlecraft, the task was defined by the 'study title' of Integral Launch and Re-entry Vehicle, or ILRV. *(NASA)*

aircraft called lifting-bodies. Shaped like flat-irons, they were designed to test ways of controlling bulbous spacecraft with stub-like wings, or convex shaped undersurfaces, providing lift during the final stage of descent through the atmosphere to a controlled landing on a runway.

Lifting-bodies were not built to go into space. They were relatively slow test vehicles, only a few going supersonic, but they all supported aerodynamic research into unconventional shapes dictated by two hybrid requirements: controlled descent at hypersonic speeds using

drag as spacecraft to slow down, and lift for controllability in the atmosphere as aircraft flown by pilots. Studies on lifting-bodies originated as far back as the early 1950s, when the heat produced by kinetic energy from friction with the atmosphere was seen as the major problem regarding flights into and out of orbit. The work paid off and coupled with tests on unmanned lifting-body shapes blasted out into space and back down through the atmosphere by converted missiles, slowly the data was gathered giving confidence in space plane technology.

Integrating roles

Rockets and converted missiles carrying objects into orbit are known as launch vehicles. Spacecraft designed to come back through the atmosphere are called re-entry vehicles. A space plane operating as a shuttlecraft would need to be an integrated launch and re-entry vehicle, essentially an all-in-one cargo vehicle carrying people and payloads back and forth to orbit. On 30 October 1968, NASA announced a requirement for aerospace manufacturers to produce proposals for such a vehicle in what it logically called its Integral Launch and Re-entry Vehicle (ILRV) study.

NASA knew it needed support for what would be a serious fight with Congress and the White House to get the project approved

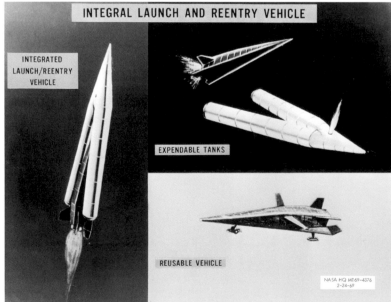

INTEGRAL LAUNCH AND REENTRY VEHICLE

INTEGRATED LAUNCH/REENTRY VEHICLE

EXPENDABLE TANKS

REUSABLE VEHICLE

NASA HQ MT69-4376
2-24-69

and it encouraged the Air Force to get behind the ILRV concept, perhaps even help fund it. NASA's budget was falling significantly each year and Richard Nixon, elected President in November 1968, was not inclined to start big spending projects when he entered the White House the following January. But in late 1968, the Air Force was playing hard to get, unconvinced that it could use an ILRV for its own military missions. So in February 1969 when NASA awarded ten-month contracts to General Dynamics/Convair, Lockheed, McDonnell Douglas and North American Rockwell, the Air Force did exactly the same, to the same companies, to see for itself if a shuttlecraft was feasible.

Then things happened fast. Sensing that the Air Force was warming to the idea, NASA decided that it would modify the ILRV studies to accommodate the much heavier payloads the Air Force would want to carry to space. Now, the vehicle would have to carry a maximum 50,000lb to orbit instead of the 25,000lb for specifically NASA-based satellites. The Air Force wanted the increased load so the shuttlecraft could lift rocket stages attached to satellites for firing into much higher orbits; NASA merely wanted to lift space station modules to low earth orbit without the necessity for boost rockets to lift the satellites, once separated from the shuttlecraft, high above earth. On 5 May 1969 Mueller told the ILRV contractors they

would need to opt for this much bigger vehicle.

The way this project was handled ushered in a new way of NASA doing business. From the outset the agency had contracted specialist manufacturers to build the space vehicles it launched. Now, with the shuttlecraft, it was introducing a phased approach: Phase A was the feasibility stage, deciding if such a concept was viable and several contractors would conduct competing studies; Phase B was the definition phase, where two contractors would compete for the contract to build the vehicles; Phase C was detailed design; Phase D was

ABOVE Emphasising the sheer scale of a future shuttlecraft, a toy airliner is held alongside a booster/ shuttle configuration to demonstrate this late-1960s concept. *(Boeing)*

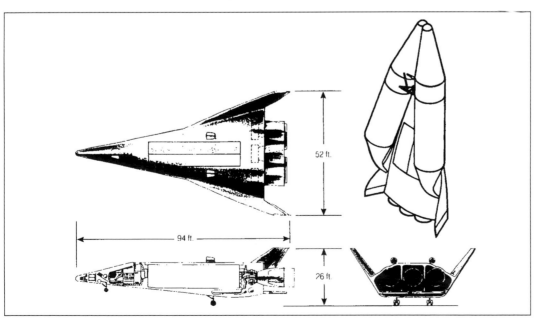

LEFT In 1968, Lockheed proposed this Star Clipper configuration, with propellant tanks forming an inverted 'V' across the front of a spaceplane. This was one concept among many proposed by aerospace manufacturers. *(Lockheed)*

DC-3, named after the world famous cargo plane of the 1930 and 1940s that served as a workhorse, lifting people and freight all around the world. It fitted the NASA image of what the shuttlecraft would do for space transportation and it was the product of a team led by Max Faget, an already legendary spacecraft designer responsible for the Mercury and Apollo capsules.

Faget applied aero-engineering logic, taking the requirement and minimising risk, reducing complexity and emphasising simplicity. He designed a two-stage vehicle comprising two rocket powered stages attached piggy-back fashion rather than in tandem. The lower stage, the booster, would propel the upper stage, the orbiter, to the edge of the atmosphere and return to a conventional landing. The orbiter would carry on into orbit under its own power. Both stages would be piloted and both would be reusable, each with relatively short, straight (low aspect ratio) wings with re-entry performed very much like the ballistic Mercury and Apollo capsules, control over the flight path only coming late in the descent.

Faget was thought by some to be over conservative and some aerodynamicists in the Air Force preferred high aspect-ratio wings,

production and operations. The ILRV studies were Phase A and it bridged the period when Richard Nixon arrived at the White House but NASA was determined not to leave it to the contractors and set about their own design.

On 23 January 1969 the NASA Manned Spacecraft Center in Houston, Texas, started work on a core shuttlecraft it wanted the ILRV contractors to work around. They called it the

creating a delta planform and thereby allowing the orbiter to be 'flown' all the way down to the ground instead of coming in pitched up at high inclination like a ballistic capsule. The Air Force disliked the lack of maneuvrability which denied the DC-3 orbiter a significant cross-range capability, defined as the ability to 'fly' left or right of the ground track during re-entry. High cross-range would give the shuttlecraft the flexibility to return to the launch site after one orbit, compensating for the westward migration of the earth's spin during the 90 minutes of a single orbit around the world.

But a delta-shaped orbiter would push technology and it would also incur greater heat loads through friction with the atmosphere. That would call for complex thermal protection materials that had yet to be demonstrated in flight. But as the ILRV studies finished and were submitted to NASA at the end of 1969, the space agency had gone through a belt-tightening process and the very nature of the shuttlecraft, and what it was designed to do, had metamorphosed into something very different. While the world had been celebrating Man's first footprints on the moon, NASA had been stripped of most of the vision that had given purpose to the shuttlecraft.

A Shuttle evolves

In September 1969 NASA presented a plan to carry it beyond the few remaining moon landings and a space station called Skylab, scheduled for launch by the last Saturn V and visited three times. Building on plans for the ILRV and permanent earth-orbiting space stations for up to 50 people at a time, NASA wanted to develop a nuclear powered rocket stage to ply between earth orbit and moon orbit, and build a series of bases, on the moon's surface and in lunar orbit, from which it would test the techniques necessary for a mission to Mars in 1981 or 1986. None of it got past the White House and by the end of 1969 NASA was uncertain that it could even get the permanent space station for which the shuttlecraft had been conceived. A partner to help share the cost, or justify the existence, of a shuttlecraft suddenly became of vital importance and when Phase B study contracts were issued to two competing teams, led by North American Rockwell and by McDonnell Douglas, in May 1970 NASA wanted it to appeal to the military to fulfil its own requirements by endorsing the Shuttle as a vital asset for the United States. But that would be hard to sell.

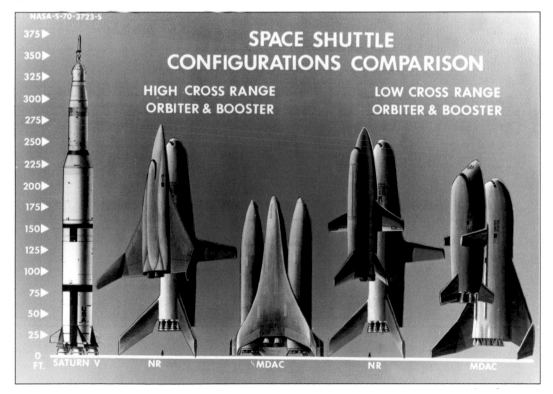

LEFT With a giant Saturn V moon rocket for scale, these high cross-range and low cross-range Shuttle configurations were proposed by the two front-running contractors, North American Rockwell (NR) and McDonnell Douglas Aircraft Corporation (MDAC). *(NASA)*

RIGHT The high cross-range orbiter proposed by North American Rockwell in 1970 would have been capable of flying up to 1,500nm either side of its orbital (ground) track. *(North American)*

BELOW The low cross-range orbiter favoured by NASA would have been smaller and lighter, with a smaller payload capability. *(North American)*

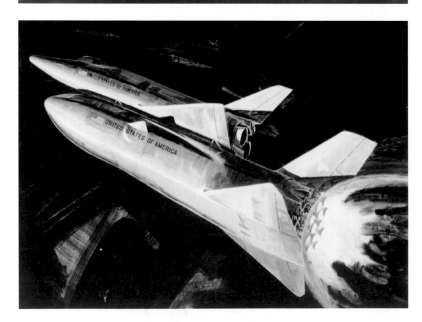

Moving to the next stage, the Phase B contracts specified a two-stage fully reusable Shuttle, examining both low (200 nautical mile) and high (1,500 nautical mile) cross-range options (cross range is a measure of how far the landing site can deviate from the orbital track), each with an orbiter payload bay 15ft in diameter and 60ft long, capable of carrying 15,000lb of cargo. But from here the specification departed from reality, anticipating up to 75 flights a year with turnaround (time for landing to re-launch) of just two weeks per orbiter, and for the whole system to be operational by the end of 1977. One of the key factors entering the picture was economics. NASA wanted the lowest possible cost-per-flight so as to encourage extensive use of the vehicle when it became operational. Because the fully reusable system was expensive to develop, annual costs spread over the seven or eight years of development could grow beyond NASA's annual budget.

As the Phase B studies progressed through the summer of 1970, NASA issued two changes. The orbiter's payload capability went up to 25,000lb and in November NASA advised the contractors that its three engines previously

LEFT As Phase B definition studies progressed through the summer of 1970, the twin vertical tail fins were deleted in favour of a single fin above the twin main engines. The booster would be powered by 12 of the same rocket motors used for the main engines. *(North American)*

thought desirable for safety reasons in case one failed was to be reduced to two. This would cut manufacturing costs, maintenance costs and turnaround time. At the end of the year NASA had set up a special programme office to give the Shuttle higher status within the agency and proposals from the two competing bidders were delivered. They were truly fantastic in every way and represented the peak of extravagant engineering and expensive design concepts.

The fully reusable system involved a 1,700 ton manned fly-back booster, bigger than a Jumbo Jet and powered by 12 big rocket engines, launched vertically carrying the 380-ton orbiter on its back. The mated stack would reach a speed of 7,000mph whereupon the booster would separate and fly back to earth leaving the orbiter powered by its two rocket engines to reach orbit and a speed of 17,500mph. The orbiter alone was more than 200ft long and as it stood on the launch pad the mated stack would be 276ft tall.

One key technology was crucial to the operation of the Shuttle – the big main engines used for the booster and the orbiter – and it was vital that these should be given a head start. To make the Shuttle work only the most advanced engine technology would do. Known as the Space Shuttle Main Engine, or SSME, it would have to be reused on more than 50 missions before replacement, be throttleable to adjust the thrust to the diminishing weight of the vehicle as it went up, and use a high pressure

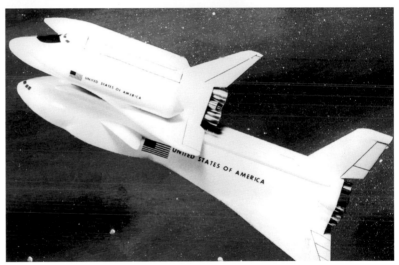

combustion chamber burning super-cold hydrogen and oxygen propellants. Moreover, the engine would incorporate pre-burners to condition the propellants (fuel and oxidizer) for maximum efficiency during combustion. Phase A studies on the SSME began back in 1968 and Phase B definition work got under way in 1970, but the challenging nature of the design brought numerous technical problems along the way.

Redefining the role

During the first half of 1971 NASA hit rock bottom. The Phase B concepts were simply too big and too costly to build and some way round this had to be found. The colossal cost of developing two separate vehicles –

RIGHT The next evolution following the over-wing hydrogen tank concept was to put the liquid oxygen beneath the orbiter in a separate tank. Boeing proposed the first stage of the Saturn V moon rocket as the booster, with the oxygen tanks above and the orbiter to one side. *(Boeing)*

FAR RIGHT An innovative proposal from Grumman resulted in the two orbiter propellants being relocated to a central tank on top of a new booster, powered by highly efficient liquid propellant rocket motors. *(Grumman)*

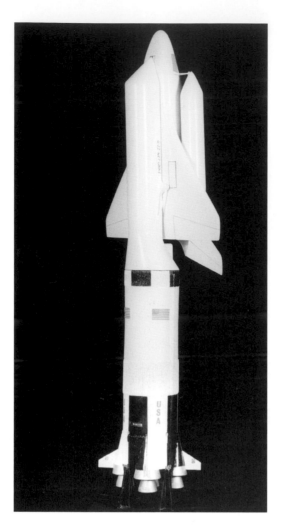

booster and orbiter – was too great for NASA to bear and in May that year NASA learned that it would have to cope with the reduced budget it already had. There would be no more money. A year earlier, in mid-1970 NASA had learned that it would not be able to afford both Shuttle and space station and that the permanently manned orbiting research facility would have to wait until the Shuttle was built and paid for. Now, on top of that, it appeared there would be insufficient money to build both booster and orbiter.

One solution was not to develop a manned fly-back booster at all, simply to reduce the size of the orbiter so that it could be launched by an expendable booster – a conventional rocket stage thrown away on each mission. It was a far cry from the original idea of a fully reusable Shuttle! Without the reusable booster the low-cost goal of cheap flights was an impossible aspiration. But even if NASA opted for a conventional booster, how could the size of the orbiter be reduced?

A solution suggested itself from the Phase A studies conducted by the Air Force. Instead of carrying all the hydrogen fuel for the two rocket motors inside the orbiter, why not carry it in two jettisonable over-wing tanks. In this way the physical size of the orbiter could be reduced because the structure for the rocket fuel could be thrown away after use. Liquid hydrogen has 75 per cent of the volume and less than 20 per cent of the weight of the two propellants, so moving it outside in separate tanks would substantially cut the size, and the cost, of the orbiter.

Since July 1970, NASA had funded studies of alternative Shuttle configurations and, in addition, it had employed a consultant – Mathematica – to work out the financial advantages in having a Shuttle at all. From the outset NASA was wedded to the concept of low-cost space transportation and found this the most useful argument in justifying the Shuttle. For every previous US manned space project, the Russian challenge to US prestige and the Space Race itself had sold the idea of putting Americans in orbit or on the moon. This time, there was no direct challenge – or reason – to build it. But without the Shuttle, US manned space flight would have been killed stone-dead at the end of the Apollo missions and the short-lived Skylab space station.

The value of the Shuttle as a potential workhorse, rather than a ferry to a space station, began to take hold. With economists looking at costs and NASA looking at technical innovation, 1971 was a seminal period in the development of its future manned space vehicle. Only by launching a lot of them could the cost per flight fall to a level that would attract its use as a satellite launcher. With the much more sophisticated technology compared to a conventional, expendable, rocket the launch price would be prohibitively expensive but with its ability to carry several satellites at once the Shuttle's flexibility and sheer carrying capacity would more than offset the high price per flight. Moreover, this price could be spread among several customers with each paying less than they would for a dedicated expendable launch vehicle.

But still the development cost was too high so NASA took the engineering a step further. What if all the propellant – hydrogen fuel and liquid oxygen oxidiser – were to be carried in a big external tank? Instead of over-wing, jettisonable tanks for the hydrogen fuel, why not a single tank carrying the two propellants in a big cylinder like a rocket stage. The winged orbiter would be attached to the side of the external tank and the tank itself fixed atop a rocket booster taking the place of the winged, fly-back, booster in the original Phase B study. It made sense and although the external tank would be thrown away after use, the simplicity and relative low cost of the tank itself would offset the price of a new one each time.

So what of the booster? It would have to be big and very powerful and a few candidates were available, from the massive S-IC first stage of the Saturn V that launched Apollo to the moon, to completely new, potentially more cost effective, rocket stages using the very latest technology to improve efficiency and performance. Boeing proposed a winged

ABOVE By late 1971, the fully manned booster/orbiter was dead, replaced by a parallel-burn booster utilising either solid or, as here, new liquid propellant motors. (*McDonnell Douglas*)

LEFT When viewgraphs showing this Grumman proposal for solid propellant boosters were first displayed at industry meetings, many warned that the unpredictability of 'solids' would be a threat to human safety. (*Grumman*)

version of the S-IC stage converting it into a reusable stage. In a series of Phase B extension studies throughout 1971, NASA gradually refined the Shuttle concept and Mathematica took each option, set it against the original concept and worked out the development cost versus the price per flight. NASA had wanted to push for the lowest cost-per-flight but that would bring the highest development cost, way out of reach. But making the Shuttle cheaper to develop meant it would be more expensive to fly. A classic Catch 22. The only way out of this dilemma was to look beyond super-efficient liquid propellant booster stages to less efficient, but very much cheaper, solid propellant rockets.

Solid propellant rockets worked like giant fireworks and they had been used effectively to improve the performance of expendable launch vehicles such as the giant Titan satellite launcher and the Delta series of middleweight workhorse rockets. But they were considered unreliable and unsafe. Once ignited, they would continue to burn until all the propellant inside had burned to depletion. If they ran amok or lost control it would be almost impossible to save the Shuttle orbiter from destruction. Until this point NASA had a rule that it would never put the lives of its astronauts at risk by flying them on 'solids'. But there were other more persuasive arguments that swept that rule

away: the cost of development and the price per flight.

Reshaping the Shuttle

By the autumn of 1971 all the teams involved in Phase B Shuttle extension studies were working hard to find a solution to the cost problem. The government had received word about the proposed development cost of the Shuttle and it was too high – way above what the Nixon administration was prepared to take to Congress for approval. By removing all the SSME propellant from inside the orbiter to an external tank, the size of the winged space plane had shrunk from over 200ft to 110ft, significantly cutting the cost of that element of the Shuttle. Now a decision had to be made about the booster, at best now only a partially reusable system, but it was not certain that NASA would have to go with solids.

Several proposals for liquid propellant boosters were around even without adopting the big S-IC stage, most of them focusing on pressure-fed liquid propellant boosters, either as a single big rocket stage on top of which would sit the external tank and the orbiter, or as two smaller rockets straddling the external tank. This was a, so-called, series-burn concept in which the booster would lift the Shuttle off

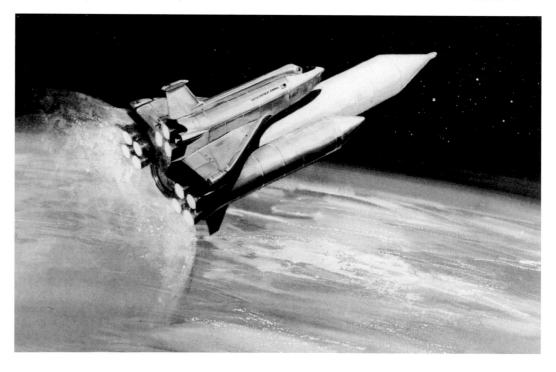

RIGHT Many at NASA, particularly at the Johnson Space Center, tried hard to find ways of developing a new generation of liquid propellant boosters with which to power the Shuttle towards space. In the configuration shown here, each booster has four nozzles and would have been reusable.
(NASA)

the pad and the orbiter would separate and ignite its engines at altitude when the booster's work was done. In the alternative concept, two smaller liquid propellant boosters would operate in a parallel-burn concept whereby the two boosters and the orbiter's SSMEs would all fire together to lift the Shuttle off the launch pad. The boosters would separate at altitude leaving the orbiter to continue on all the way to orbit, only separating the external tank when the SSMEs shut down.

With series-burn, only a big liquid propellant rocket stage would do because solids of the required thrust had only ever been used in tests in the early 1960s. With the complete Shuttle weighing almost 2,200 tons fully fuelled, the booster would need to produce a thrust of 3,300 tons to get off the pad. With the series burn concept all that thrust would have to be provided by the booster alone. With the parallel-burn approach, however, the orbiter's SSMEs would contribute approximately 700 tons of thrust, leaving the two solid boosters to each have a thrust of only 1,300 tons instead of more than double that for a single boost stage – and they were cheaper too.

So, by the end of 1971, the Shuttle concept had been refined down to a preference for a winged orbiter with a 60ft x 15ft payload bay, an external tank carrying the hydrogen and oxygen for the orbiter's SSMEs, and a parallel-burn booster concept using either two pressure-fed liquid propellant motors or two solids. In either case, the plan was to recover the booster from an ocean splashdown and use them again. But there was a further impediment to stripping costs even further, and the reason went right back to when the Air Force was encouraged to get behind the Shuttle as a concept and support its development in Congressional budget hearings.

If the Air Force was to use the Shuttle it would have a big say in the orbiter's payload capacity. NASA had wanted a Shuttle capable of putting 15,000lb of satellites or cargo in low orbit, about 200 miles above earth out of Cape Canaveral in Florida. The Air Force wanted to place 40,000lb in a polar orbit from Vandenberg Air Force Base in California. But the increase was more than that implied by the numbers.

NASA's mission would take it due east on a path that would take advantage of the spin of

ABOVE Cost and the unassailable logic of economics forced NASA to adopt solid propellant boosters when it had been said that the space agency would never fly people on that type of rocket motor. *(Grumman)*

BELOW From a fully reusable system to a twin solid rocket motor concept, between summer 1970 and spring 1972 the Shuttle was drastically redesigned to enable a lower development cost. *(NASA)*

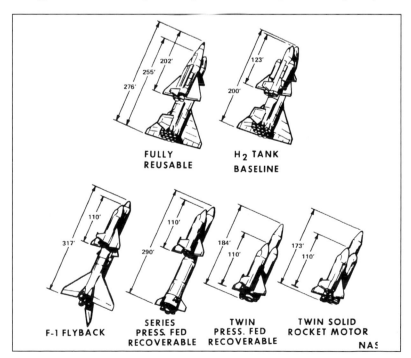

EARTH ORBIT CARGO TRANSFER

MSFC-70-PD-4000-25B

the earth as it rotated on its polar axis. The Air Force wanted to send satellites into space that would head due south and circle the earth at 90 degrees to the equator, passing over both poles. That meant the energy needed to get it into orbit had to be that much greater because it could not rely on the earth's spin to add more than 800mph to the launch speed. This meant that 40,000lb to polar orbit was equivalent to 65,000lb out of Cape Canaveral, more than four times NASA reference mission of 15,000lb to a due-east low earth orbit first proposed in 1970 at the start of Phase B studies. But it was a price that had to be paid because accommodating the Air Force was an essential part of getting the Shuttle approved.

It would be wrong to lay the blame for payload growth wholly upon the Air Force. Because eventually NASA wanted to assemble a space station from separately launched modules, it too had a requirement for a 45,000lb payload because, as they would be placed in orbits inclined 55 degrees to the equator that would equate to a 29,000lb due-east capability. And there was another reason for a bigger lift. Both NASA and the Air Force wanted to develop what they called a Space Tug, a rocket stage that would push satellites and spacecraft into higher orbits than the Shuttle could reach, and that too would have to be lifted up along with the payload.

But the price for the Air Force missions was to build a technically complex Shuttle with 1,500 nautical mile cross-range capability that could allow the orbiter to 'fly' back to its launch site after one full orbit of the earth because one mission the Air Force was not about to publicise was the one in which it wanted to refuel in space the next generation of spy satellites. Big satellites, each the size of a bus, that could virtually read the

headlines of a newspaper from orbit and which would need to remain in space for many years to maximise the value of its cost. That drove the technology for the orbiter by requiring a delta wing to make the high-lift turns during re-entry, and that in turn meant a more sophisticated – and in the end fragile – thermal protection system that would one day destroy an orbiter and take the lives of seven NASA astronauts.

ABOVE NASA boasted of the Shuttle's value as a cargo lifter, with a Space Tug carrying payloads and people off to the moon and high-orbiting space stations. *(NASA)*

BELOW Although it was never developed, the Shuttle-launched Tug was to have been an integral element in the National Space Transportation System, ferrying people and cargo around in space. *(NASA)*

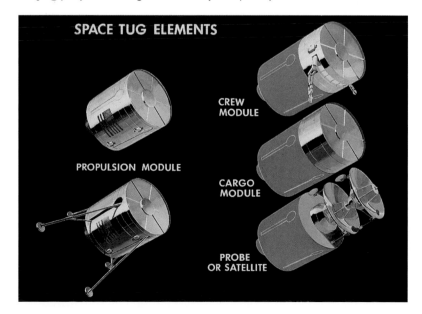

SPACE TUG ELEMENTS

PROPULSION MODULE

CREW MODULE

CARGO MODULE

PROBE OR SATELLITE

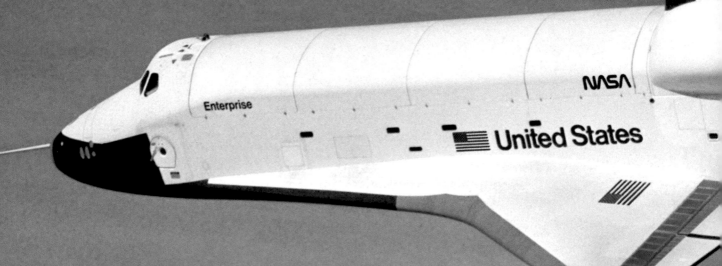

Chapter Two

The Shuttle flies

Delays and technical obstacles provided challenges to NASA *en route* to getting the Shuttle flying. It finally rocketed into space on 12 April 1981 – 5 years later than originally planned and 20 years to the day after Yuri Gagarin became the first human to orbit the earth. Before that first historic Shuttle flight lay a host of tests and preparations, only a handful of which had been anticipated at the outset. It was to be a long road, with many pitfalls along the way.

LEFT Fitted with a tail cone to reduce drag, Shuttle Orbiter OV-101 *Enterprise* glides down to a safe landing at NASA's Dryden Flight Research Center, California, during the second of five air-launched drop tests from the top of an adapted Boeing 747 Shuttle Carrier Aircraft in 1977. *(NASA)*

By the end of 1971 the Shuttle had reached a point where it was either going to be accepted by the government or cancelled outright. On 3 January 1972, NASA boss James Fletcher met with George Schultz, director of the Office of Management and Budget, the OMB that held the nation's purse strings, and gave the Shuttle its blessing. Two days later Fletcher flew out to the Western White House at San Clemente, California, for a brief meeting with President Nixon, whose agreement to approve the Shuttle was essential to its next step – acceptance by Congress and the budget to make it happen. It was already a done deal and the meeting was a photo-call, with Fletcher taking along a model of the Shuttle that caught Nixon's imagination and left Presidential aide John Erlichman wondering if the NASA boss was ever going to get it back!

NASA had done a good job on the President, enthusing him with visions of ordinary citizens traveling into space, of all manner of satellites and spacecraft launched into orbit by the Shuttle, and of cheap access to orbit from where the next generation of explorers might leave for new destinations in the solar system and where a permanently manned orbiting space station would be assembled – but not

RIGHT For both the proposed Pressure Fed Booster and Solid Rocket Motor concepts, abort Solid Rocket Motors (SRMs) were considered essential for crew safety should anything go wrong during ascent. *(NASA)*

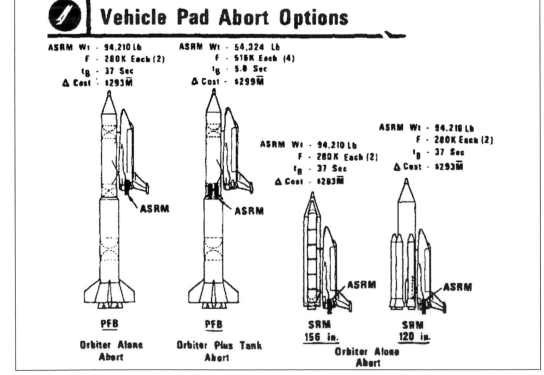

yet. There were still the hurdles of final selection about the precise engineering configuration of the Shuttle – liquid propellants or solid rocket boosters; series-burn like a rocket, parallel-burn with all engines firing at lift-off.

And then there was Congress. Would they approve the money, approval without which all the enthusiasm in the world would not get the Shuttle built? It was a comparatively easy ride with the legislature, the executive White House, having put their shoulder behind the project Washington politicians were largely supportive, believing NASA's optimistic assertions about low-cost space travel. That would only happen if NASA launched a lot of flights and by early 1972 it had staked its reputation on an ambitious manifest which envisaged 580 missions over a 12-year period starting in 1979, when the Shuttle was now expected to start flying.

Seen with 40 years of hindsight, it is extraordinary that nobody seriously challenged that figure. But averaging almost 50 flights a year, it would indeed prove cheaper than an equivalent space programme using expendable rockets. Of the total 580, 62 per cent of all flights were allocated equally between the military and NASA flights supporting space

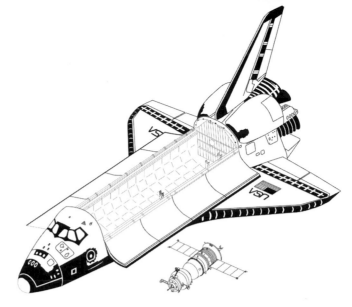

station logistics missions and sending up commercial satellites. The remaining 38 per cent were to have been scientific research flights using the interior of the Shuttle or a laboratory module fitted to the inside of the payload bay.

In approving the Shuttle, Congress stipulated that it must cost no more than $5.5bn to develop, a price NASA said it could work to because by choosing solid rocket boosters it would save $700m over the liquid propellant candidates and stay within that figure. That eliminated the liquid rocket boosters as candidates and in March 1972 NASA issued requests for proposals to enter Phase C/D

ABOVE Compared to the Russian Soyuz spacecraft (pictured beneath the Shuttle), the Shuttle Orbiter was a giant leap in technology. The Soviet propaganda machine claimed that the Shuttle's large cargo bay could be used for capturing Russian spacecraft! (NASA)

LEFT The definitive concept involved the firing of the Orbiter's three main engines and the two solid rocket boosters for lift-off and ascent towards orbit, a so-called parallel-burn approach.
(North American)

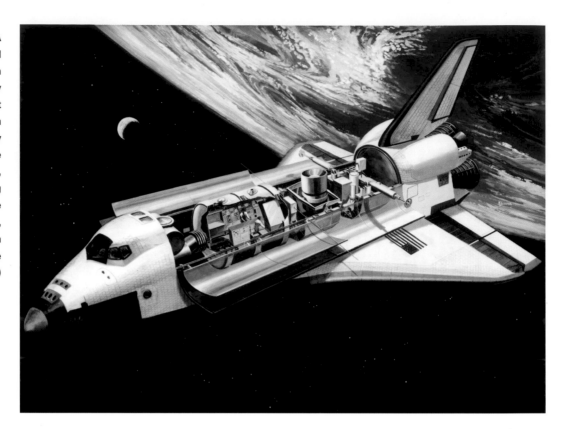

of detailed design and manufacturer with a single contractor building the orbiter. Other contractors would be selected later for the external tank and the solid boosters. Rocketdyne had already been selected, in July 1971, to build the main engines, each orbiter now planned to have three fitted in the aft fuselage, upped from an earlier decision to fit two because more thrust would be needed for the parallel-burn mode.

Four contractors (a Grumman/Boeing team, Lockheed, McDonnell Douglas, and North American Rockwell) bid for the prestigious job of building the orbiter and integrating the other elements into a fully flyable Shuttle. When the bids were scrutinised, Lockheed and McDonnell Douglas were eliminated and when NAR was shown to have the lowest bid, the choice was clear. On the evening of 26 July 1972, NASA announced that North American Rockwell would build the Shuttle Orbiter. The contract to build the External Tank (ET) went to Martin Marietta on 16 August 1973 and on 20 November the Solid Rocket Booster (SRB) contract went to Thiokol, with a division of United Technologies awarded a contract on 21 December 1973, to fabricate the non-motor

elements of the SRB, such as the nose cap, recovery parachutes, skirt assembly, etc.

An evolving design

While the Shuttle selected for contract in 1972 looked superficially very close to the Shuttle that first flew on 12 April 1981, the first two years after contract award saw a multitude of changes to the detailed design and to the way in which it would operate. Weight is critical to the performance of any launch system and the Shuttle was truly an integrated launch and re-entry vehicle, as had been defined by the ILRV studies several years earlier.

Because it was designed to fly back down through the atmosphere and land on a 10,000ft runway, engineers wanted it to have conventional turbofan engines which would pop out from a compartment at the rear of the payload bay and give the pilots power to go around again if the approach was off track or if something prevented them landing immediately. This 'go-around' capability imposed too great a penalty in weight and volume for it to remain. So the proposed engines were deleted from the design, relegating the Orbiter to a 100-ton glider

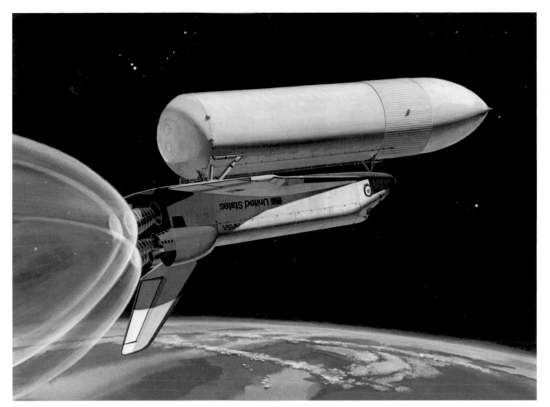

LEFT On early flights the Orbiter would hang beneath the External Tank (ET) until the ET was separated and the Orbiter rolled upright, but on later flights the Shuttle would roll heads-up during the terminal phase of ascent to enable communications through the Ku-band antenna. *(North American)*

LEFT Just discernible above the wing, and bearing a tapered nose section, is the abort motor which, paired with a second on the other side, would have carried the Orbiter to safety in the event of a malfunction during ascent. *(NASA)*

without any means of propulsion of its own once it entered the earth's atmosphere.

The critical part of the launch process was the two minutes that the SRBs were firing. If something went awry with the boosters during this time engineers predicted that the forces on the Orbiter could be catastrophic. They were. In January 1986 the Shuttle *Challenger* was destroyed by a failed booster. The initial design requirement back in 1972 had been to have blow-out ports at the forward end of each booster so that the flow of hot gases could be diverted through the front ports, slowing the entire assembly, known as 'the stack', allowing the Orbiter to separate and fly back down to a controlled landing. This too was removed to save weight.

Crew escape from a catastrophic failure during launch occupied the minds of engineers for months and a wide range of possibilities were studied, including the entire crew cabin being a jettisonable escape pod. The complexities and excess weight of such a design were deemed prohibitive. But to get the Orbiter off the big SRBs if they ran amok, engineers proposed the use of Abort Solid Rocket Motors, ASRMs. Attached each side of the rear fuselage adjacent the vertical tail, the two ASRMs would fire to carry the Orbiter away from the External Tank. If they were not needed during ascent, the weight penalty they incurred could be compensated by firing them to add thrust to the Orbiter's SSMEs after the SRBs stopped firing and separated. But these too were dropped for a simpler system – a belief that the Shuttle could be made so safe that its reliability would be as high as that of a commercial airliner.

The naive assumption that the Shuttle would be a fully operational vehicle after only a few test flights was all pervading and brought about a belief that space systems were sufficiently advanced that these multiple safety systems were unnecessary. In essence, choosing technical engineering solutions embedded throughout the system over physical safety and abort systems. But the Shuttle was an entirely new concept. It was the first of its kind and there was no precedent upon which to base its design; and it was one of a kind and therefore had no parallel from which to draw lessons. And

that over-simplistic view, that the Shuttle could be made super-safe, wrote a new mathematical logic: fail-operational/fail-operational/fail-safe. Meaning, it would be designed so that no one system would fail the mission for which it had been launched, that no multiple of such failures could threaten the ability of the Orbiter to return at will, and that no systems or structural failure would threaten the lives of the crew. The mathematics forecast a reliability level such that NASA expected to fly more than 1,000 missions before losing a crew. In reality, it lost two crews in little more than 100 flights.

A time to build

The Shuttle contract was for four Orbiters – one to be used for drop-tests off the back of a Boeing 747 and three for space flight, the first one being refurbished and to join the space-fleet later. There was a need to move the Shuttle around the country driven by the decision to have two launch sites: the Kennedy Space Center at Cape Canaveral run by NASA, and the Vandenberg Air Force Base in California for the Air Force polar-orbit flights. The West Coast launch site was necessary because to fly north into polar orbit from Florida would carry the ascent over populated areas, which was prohibited. So the Air Force would launch south from Vandenberg over water. Also, Orbiters would have to be moved between sites and from emergency landing strips if they had to come down at one or the other for bad weather avoidance at the preferred landing strip. Also, in an emergency the Orbiter might have to land in a foreign country and the Boeing 747 would be needed to fly it back.

The aircraft selected for the role of Shuttle Carrier Aircraft, or SCA, was an ex-American Airlines 747-100 (N9668), purchased on 18 July 1974, and modified to carry Orbiters horizontally on struts using attachment points usually used when it was connected to the External Tank. The most noticeable difference was in the small vertical fins attached to the extremities of each horizontal tailplane, with special damper struts located inside the forward fuselage. NASA purchased a second 747-100 (JA8117, re-registered N911NA) in April 1988. To prevent excessive drag caused by the blunt aft end of the orbiter, a special tail cone was built by

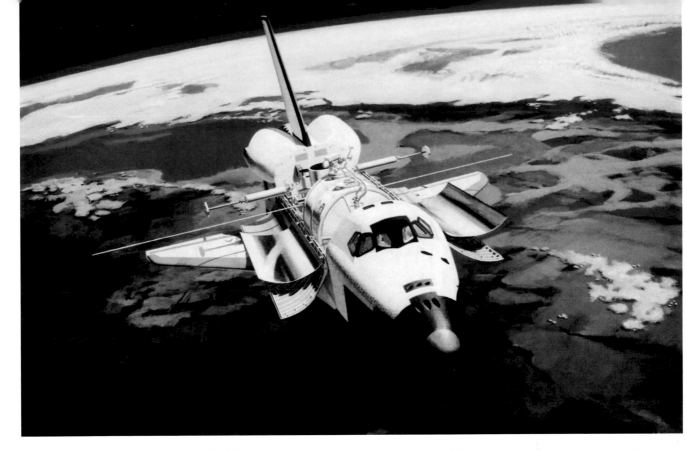

Boeing, 36ft long, 25ft wide and 22ft high, covering the space normally occupied by the three main engines, the SSMEs.

Because the Shuttle was unique, astronauts would have to learn how to land this 100-ton glider. Not only because it had no engines of its own but because of the unusual approach angle during the final stages of descent toward the runway. Whereas most airliners approach the runway at an angle of 3–5 degrees to horizontal, the Orbiters would dive down upon the runway at an angle in excess of 20 degrees, only pulling up just a few hundred feet above the ground at the flare point. Simulators could

ABOVE The Orbiter was designed for many tasks, and the cargo bay was considered a platform from which a variety of scientific instruments could operate. *(North American)*

LEFT With tail cone attached, Orbiter *Discovery* sits atop its Boeing 747 Shuttle Carrier Aircraft at Dryden Flight Research Center, California. *(NASA)*

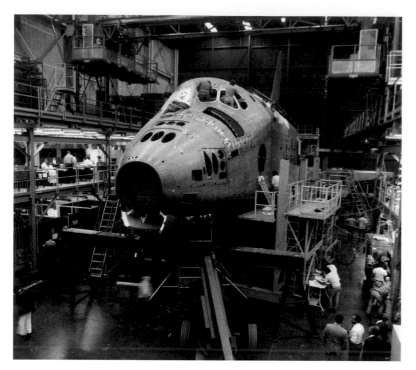

ABOVE **Shuttle Orbiter STA-099 was built for structural load testing, but would be rebuilt into OV-099 and named *Challenger*.** *(NASA)*

reversal in the air, usually applied as a braking function after touchdown, and differential deployment of main and nose landing gear, so that the main wheels alone could be deployed for maximum drag and give the aircraft the necessary high-drag descent profile. In the cockpit the left side was a close copy of a Shuttle Orbiter and the usual control yoke was replaced with a central stick-controller, as carried by the Orbiter. On the right side, it was a standard Gulfstream II.

By the time assembly of the first Orbiter was complete in 1975, at government and academic facilities across the United States engineers had spent 46,000 hours finessing the definitive shape and outline of a truly unique flying machine. When the contract had been awarded in July 1972 the design configuration was known as Vehicle 1 and within two years the definitive Vehicle 6 had been sealed as the shape of the Shuttle to build. Many factors entered into the subtle shift in Orbiter design, the precise arrangement of External Tank and Solid Rocket Boosters and the way the vehicle would fly and operate. Everything fed back into a blend between extrapolated data from limited experience with the hypersonic X-15 and complete unknowns catered for purely through calculation and scale tests.

Two engineering tools were important to evaluating the system: the Main Propulsion Test Article (MPTA-098) and the Structural Test Article (STA-099). The MPTA would be used to test the main engines fitted to an aft fuselage structure that simulated the back end of an Orbiter. Set vertical in a test rig at the National Space Technology Center (NSTL), Alabama, until 1973 known as the Mississippi Test Facility previously used to test rocket engines for the Saturn rockets. In 1989 it would be renamed again, currently known as the Stennis Space Center. The first tests with a Shuttle main engine (SSME) took place in July 1975.

Meanwhile, assembly of STA-099 began on 21 November 1975. Late in 1977 a decision was made not to plan on modifying *Enterprise*

familiarise pilots with the techniques but only by flying an aircraft that could simulate this approach could full confidence be acquired.

NASA bought two Grumman Gulfstream II business jets (N946NA and N947NA) and modified them to be capable of flying Shuttle-like profiles, simulating a Shuttle descending through the last 35,000ft of its descent. Known as Shuttle Training Aircraft (STA), they were modified with the ability to command thrust-

LEFT **STA-099 in its enclosure for load and stress analysis, which involved using jacks and weights to simulate the more demanding conditions of ascent and re-entry.** *(North American)*

(OV-101) for space flight after its air-drop tests but rather to take STA-099 and upgrade it for joining the space fleet as OV-099 named *Challenger*. Assembly of STA-099 was completed in early 1978 with rollout on 18 February but its primary task was to test the stresses of varies loads applied across its structure. To do that it was wheeled across from Plant 42 at Palmdale to Lockheed's facility for stress tests involving a classic loads analysis in a special test rig containing 256 jacks inside a steel rig weighing 430 tons. Loads applied to 836 different locations were carefully measured, stresses duplicating conditions the Orbiter would experience during launch, ascent into orbit, re-entry and landing. Because a decision had been made to adapt the Orbiter for space flight, and fearing the Orbiter could experience some damage, loads to only 120 per cent of calculated conditions were applied rather than 140 per cent which is the industry standard for airframes.

A time to fly

The first flight-rated Orbiter, OV-101, was originally to have been named *Constitution*. Due to a public letter-writing campaign the name was changed to *Enterprise*, from the Starship in the TV series *Star Trek*. When rolled out on 17 September 1976, it was a seminal moment for the entire Shuttle programme

and Gene Roddenberry and most of the *Star Trek* cast were on hand at Rockwell's plant at Palmdale, California, to help the nation celebrate. But *Enterprise* had no provision for main engines, no thermal protection and was incapable of flying into space but would be used instead for a series of drop tests from the Boeing 747 SCA. Nevertheless, it was the first Shuttle and would be the first to fly freely within the atmosphere.

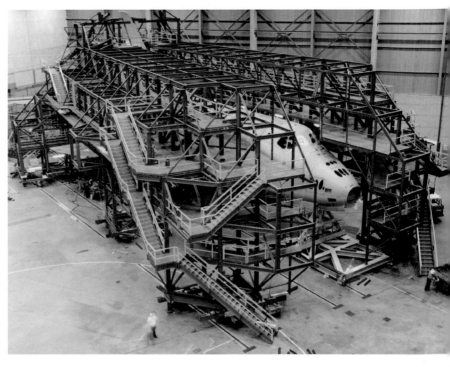

ABOVE The simulation of conditions that the Orbiter would experience was based on careful analysis, but flight-test results would highlight a few unexpected results. *(North American)*

LEFT NASA's M2-F1 Lifting Body, a concept that originated at NASA's predecessor organisation, the NACA, in the 1950s, and provided much aerodynamic research, gives dramatic scale to OV-101 *Enterprise*. *(NASA)*

Enterprise was taken 36 miles by road from Plant 42 at Palmdale to the Dryden Flight Research Center, a NASA flight test facility on part of the famous Edwards Air Force Base from where Chuck Yeager had flown through the sound barrier and from where all the X-planes, including the hypersonic X-15, had made their record breaking flights. It arrived on 31 January 1977, for a flight-test programme that would last for much of that year. The plan was to place *Enterprise* with its tail cone attached on top of the Boeing 747 SCA and to test the flying and handling characteristics, followed by a further series of flights with powered applied to the Orbiter and two pilots on its flight deck. Only then would *Enterprise* be flown off the top of the SCA for a free flight down to a runway at Edwards Air Force Base.

Taxi tests on 15 February were followed three days later by the first of five 'captive-inert' flights to test handling. Three 'captive-active' flights began on 18 June with astronauts Fred W. Haise (of Apollo 13 fame) and C. Gordon Fullerton on the Orbiter's flight deck. Finally, on 12 August, the first of five drop tests took place. With Haise and Fullerton piloting *Enterprise*, the Orbiter lifted free as the Boeing SCA went

into a very shallow pushover of 7 degrees. Released at an altitude of 24,100ft and a speed of 310mph, the crew piloted the *Enterprise* off the top of the SCA and down to a safe landing 5 minutes 21 seconds later, touching down at a speed of 213mph and rolling to a stop in 11,000ft.

Two more free flights were made, on September 13 and 23, with the tail cone attached to the aft end of *Enterprise*. To test the true configuration of an Orbiter returning from space, the tail cone was removed for the fourth free flight on October 12. By removing the tail cone the Orbiter imposed much greater buffeting on the Boeing 747 and, with much greater drag, would descend to earth in less than half the time with the tail cone on. Released at the lower speed of 290mph and a height of 22,400ft, the Orbiter returned to earth in just 2 minutes 34 seconds, plunging to earth like a brick and giving astronauts Joe Engle and Richard Truly a thoroughly realistic rehearsal for returning from space. A fifth and final free flight, again with the tail cone removed, gave Haise and Fullerton a chance to experience the exciting ride which, falling at more than 9,000ft a minute was more akin to a fast fall in an elevator.

RIGHT NASA Administrator James Fletcher explains the nuances of the Shuttle to various members of the *Star Trek* cast attending the rollout of *Enterprise* at Palmdale, California, on September 17, 1976. *(NASA)*

ABOVE Pilots reported the handling of the Boeing 747 Shuttle Carrier Aircraft (SCA) was little compromised by the presence of the Orbiter on top. *(NASA)*

LEFT With the SCA going into a gentle dive, *Enterprise* 'flies' off the upper fuselage fixtures, similar to those that will secure its successor to an External Tank for space flight. *(NASA)*

A time to get ready

With free-flight tests of *Enterprise* clearing the Shuttle for that part of its mission, on 13 March 1978 the Orbiter was flown on the back of the Boeing 747 halfway across the country, from Dryden in California to the NASA Marshall Space Flight Center (MSFC) in Alabama. It was there for shake tests in what NASA termed the Mated Vertical Ground Vibration Test (MVGVT) facility, a large building looking like a square-shaped grain silo in which it would be exposed to punishing vibrations. In there it would experience the violent conditions of engine ignition and launch, it would be attached to a dummy External Tank and shaken as it would be during ascent, and dummy Solid Rocket Boosters would be attached for duplicating the pounding an Orbiter would get from sitting between two sets of solid rocket motors each delivering more than 80 million horsepower of energy!

When that work was done, in April 1979 *Enterprise* was again hoisted aboard the Boeing 747 and flown to the Kennedy Space Center. Rolled round to the Vehicle Assembly Building (VAB) which had been built in the mid-1960s for assembling Apollo-Saturn rockets, it was attached to the assembled External Tank slated for the first manned mission, and two inert SRBs. On 1 May 1979, spectators got their first sight of a fully stacked Shuttle being rolled out to Launch Complex 39A, the pad from where Apollo flights had been launched a decade earlier. For three months it remained at the pad verifying the correct location of maintenance work platforms.

On 23 July 1979, the stack was rolled back to the VAB and on 15 August *Enterprise* was flown to Vandenberg Air Force Base where workers building the Shuttle launch pad at that site could get their first view of a Shuttle. From there it was flown the short distance to NASA's Dryden Flight Research Center and on October 30, 1979, it was trucked overland back to Palmdale where it had been built. Its main job was over but it would be used again to herald the coming age of reusable space transportation. In May 1983, *Enterprise* was again attached to the Boeing SCA and flown to the Paris Air Show, wowing crowds. In April

ABOVE *Enterprise* performed a variety of roles, one of the most important being at the Mated Vertical Ground Vibration Test (MVGVT) facility in Alabama, where it would undergo vibration and shake tests. *(NASA)*

RIGHT Hoisted by a sling similar to that used in the Vehicle Assembly Building at the Kennedy Space Center, *Enterprise* is raised into position in the MVGVT building. *(NASA)*

the following year it was a feature exhibit at the World's Fair in New Orleans and in late 1984 and early 1985 it was used to check out the specially built launch pad at Vandenberg Air Force Base, SLC-6 (Space Launch Complex No.6), much as it had at the Kennedy Space Center five years earlier.

Meanwhile, development of the Space Shuttle Main Engine had been moving apace with tests undertaken by the prime contractor Rocketdyne, which in 1984 was to become a division of Rockwell International (formerly North American Rockwell), the builder of the Shuttle orbiters. Managed by the Marshall Space Flight Center the engines were a technological challenge pushing motor technology to the known limits in the 1970s. Development of the SSME was a hard, tough struggle sandwiching design innovation between schedule pressure from NASA and engine failures in tests.

Gradually the complex and high demanding design brought positive results and accumulation of successful test time began to stack up. NASA's Shuttle programme manager John Yardley brought aircraft engineering practices from his former post

LEFT Checking the structural fit of launch facilities and Orbiter at the Kennedy Space Center, *Enterprise* is supported by the External Tank assigned to the first flight plus empty boosters. *(NASA)*

BELOW Checking another facility on the West Coast, *Enterprise* is pictured at SLC-6 at Vandenberg Air Force Base, California, built but never used for the Shuttle. *(USAF)*

ABOVE The forward section of the Shuttle Solid Rocket Booster is stacked in the Vehicle Assembly Building at the Kennedy Space Center. *(NASA)*

ABOVE The Solid Rocket Motors used for the Titan launch vehicle proved reliable and effective, technology from which much was drawn for Shuttle operations. *(USAF)*

RIGHT The External Tank is rolled out, displaying the aft attachment for the Orbiter, and the external liquid oxygen line extending down the side of the liquid hydrogen tank. *(Martin Marietta)*

at McDonnell Douglas to the job and set a baseline requirement of 65,000 seconds of engine testing as the criteria for clearance to fly. He achieved that number thus. In aircraft flight testing, 40 consecutive successful flights of a new engine would qualify it for clearance. Because each engine was required to operate for 520 seconds to get the Shuttle into orbit, he multiplied that figure by 40, getting 20,800 seconds. And because the Shuttle would carry three of those engines, he multiplied that figure again by three. On 24 March 1980, the SSME achieved more than 65,000 seconds and by the time the first Shuttle was launched a year later total successful engine time reached 110,253 seconds in 726 engine firings.

Throughout this period the Solid Rocket Boosters had been under development by Thiokol. Based in Utah, the company had produced the world's biggest solid propellant rocket motor and it had achieved that based on little precedent. Research into big solid rockets had been going on since the early 1960s but nothing on the scale of the Shuttle's SRBs had been attempted. The Titan missile, adapted into a satellite launcher as the Titan 34D employed two solids, each producing a thrust of approximately 600 tons, less than half the thrust of a Shuttle SRB. Four SRB development motor firings were conducted by Thiokol between July 1977 and February 1979 to gather performance data, followed by three qualification firings between June 1979 and January 1980 to certify the motors for flight.

Least problematical was the External Tank, the central structure flanked by two SRBs and to which would be attached the Shuttle Orbiter. In what looked strongly similar to a Saturn rocket stage, the Shuttle ET was a marvel of engineering and loads analysis. With the hydrogen tank at the bottom and the oxygen tank at the top, reversed from the position on Saturn stages, when fuelled the tank was the heaviest single element of the stack and weight was always going to be a target for trimming. Manufactured by Martin Marietta in St Louis, elements of the tank's design were tested together in March 1977 in a special rig at the Marshall Space Flight Center. Because the super-cold (cryogenic) propellants needed insulating from the ambient atmospheric

temperature, a special insulation was developed and this was tested in special chambers. MSFC also hosted structural loads tests – this time to the full 140 per cent recommended by industry. Tanks could only be used once. A major programme of weight reduction began long before the first flight although it would be several years before these savings were introduced onto the production line.

A time to launch

Assembly of the first Orbiter destined for space flight began on March 27, 1975, less than 14 years after Yuri Gagarin became the first human in space. Named *Columbia* after a sloop based in Boston, Massachusetts, famous for making the first American circumnavigation of the globe, OV-102 was finally rolled out on 8 March 1979. *Columbia* was also the name chosen for the Apollo

ABOVE The External Tank was designed and developed in cooperation with the NASA Marshall Space Flight Center, where it was subjected to a wide range of pressure and vibration tests. *(NASA)*

spacecraft carrying the first moon landing crew to the lunar surface in July 1969. With the Shuttle programme already running a year or so late, there was pressure on NASA to begin flights but there were insidious flaws in the system that would bring about a succession of delays. Succumbing to pressure to get the Shuttle into space, *Columbia* arrived at the Kennedy Space Center atop its Boeing 747 SCA on 25 March 1979. Few could have

foreseen that it would still be sitting there two years later.

With development problems on the main engines, NASA now encountered trouble with the worrisome thermal protection system comprising more than 30,000 separate ceramic tiles designed to protect the undersurface and sides of the Orbiter from the heat of re-entry. Manufactured for resilience and reusability, tests in 1979 revealed that the tiles had inadequate

tensile strength and there were problems too with the bonding adhesive which had to keep all the tiles in place. In some areas they would protect the Orbiter's vulnerable aluminium structure from temperatures as high as 2,300 deg F.

Various techniques were tried and all appeared resistant to solutions but gradually, over time, the problems were resolved. *Columbia* arrived at KSC with 24,000 tiles installed and 6,000 to be added in the Orbiter Processing Facility (OPF) where it was prepared for flight. But the tile tests revealed that almost all the tiles fitted in Palmdale would have to be replaced – some had even fallen off flying it across country – and for 20 months a veritable army of recruits worked three shifts a day, six days a week, installing a total of 30,759 tiles.

The External Tank for the first Shuttle mission, STS-1 (Space Transportation System-1) arrived at the Cape on 29 June 1979, and was mated with the two SRBs destined for this historic event on 3 November 1980. *Columbia* was mated to the ET on 26

LEFT Preparing *Columbia* for flight was a complex and time-consuming endeavour, made worse by a steep learning curve on tile application! *(NASA)*

BELOW *Columbia* sits on Launch Complex 39A, as two converted Grumman Gulfstream II Shuttle Training Aircraft overfly the hard stands prior to launch. *(NASA)*

ABOVE Shuttle astronauts John Young (foreground) and Robert 'Bob' Crippen practice operations in the aft flight deck area on the Orbiter simulator. *(NASA)*

BELOW *Columbia* rolls to the launch pad on 29 December, 1980. Only the first two tanks carried the white fire-retardant latex coating. *(NASA)*

November after spending a record 613 days in the OPF. Finally, the stack was rolled out to the pad on 29 December 1980. Because the flight engines for *Columbia* had not been fired as a group within that Orbiter, NASA wanted to give them a quick firing test on the pad, so a Flight Readiness Firing (FRF) took place on 20 February 1981, when engineers 'blipped the throttle' for 20 seconds to see if it all held together, including the tiles. Each new Orbiter would conduct a FRF.

On 19 March 1981, the flight crew comprising veteran astronaut John W. Young and Robert L. Crippen were on the flight deck for a scheduled test when two technicians working in the main engine compartment at the rear of the Orbiter were asphyxiated by nitrogen during a purge of that chamber. Both men lost their lives. Numerous tests were conducted, including loading and unloading the vast External Tank, and all appeared ready for a flight on 10 April when a problem between timing systems in the avionics caused a scrub in the launch attempt for that day. Two days later, on 12 April 1981, exactly 20 years after the flight of the first human into space, *Columbia* thundered into life and the world's first reusable Shuttle was launched.

Although every element of the Shuttle had been tested separately, this was the first time astronauts would fly in a vehicle that had not already been first tested in space. Previous manned spacecraft (Mercury, Gemini and Apollo) had all been flown unmanned several times before astronauts were allowed to fly in them. With the Shuttle that was not feasible because it was both a launch vehicle and a spacecraft and it was not designed to fly unmanned. In that regard it was like an aircraft which had to have a pilot to fly. To have fitted it out with an unmanned capability would have significantly delayed the programme and not represented the configuration capable of carrying astronauts.

The FRF had demonstrated a 'twang' effect, where the offset alignment of the Orbiter's engines causes the entire vehicle to bend over about 20 inches in response to the shock of ignition. It was necessary to allow the stack to reflex back before ignition of the SRBs. The countdown reached zero at ignition of the

ABOVE **Illuminated by pad lights, *Columbia* spent 103 days on LC-39A prior to its launch.** *(NASA)*

Orbiter's three SSMEs, followed four seconds later by ignition of the two big SRBs and lift-off. On later flights the countdown would incorporate ignition of the SSMEs within the final few seconds so that lift-off came at T-0. As a result of engineering analysis on STS-1, the duration of the burn prior to ignition of the SRBs would be extended to allow more time to null the 'twang' effect.

At lift-off the Shuttle weighed 2,200 tons (4,457,111lb) and the two SRBs burned for 2 minutes 10 seconds carrying the stack to a height of 31 miles. Separating from the External Tank they fell back into the Atlantic Ocean

RIGHT **The world's first reusable Shuttle ascends from LC-39A at 07.00hrs local time on 12 April, 1981, to begin mission STS-1, the first of more than 130 flights.** *(NASA)*

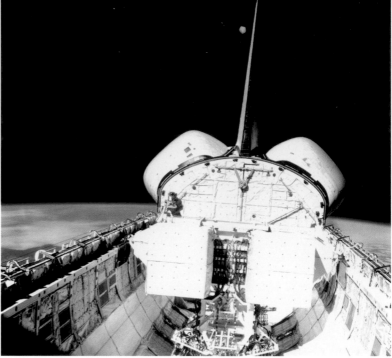

supported by parachutes, impacting the water at 7 minutes 10 seconds, 161 miles downrange from the Kennedy Space Center. The SRBs were recovered by a special vessel and towed to Port Canaveral from where the booster segments would be cleaned out and used again. The remaining Orbiter and External Tank now weighed just over 700 tons (1,476,278lb) and the Shuttle arched over on its way into orbit burning cryogenic propellants from the ET. As the three main engines consumed propellant and made the stack

lighter, they were gradually throttled back to contain acceleration within 3g.

The computers shut down the three SSMEs at 8 minutes 34.4 seconds. Having achieved altitude, in the final few minutes the Shuttle had been flying slightly down toward the earth, gathering speed and at separation of the ET, 24 seconds later, the Shuttle was at an altitude of 72½ miles and not quite at orbital velocity. The External Tank gradually descended into the earth's atmosphere and broke up at an altitude of 54 miles over the Indian Ocean, most of it

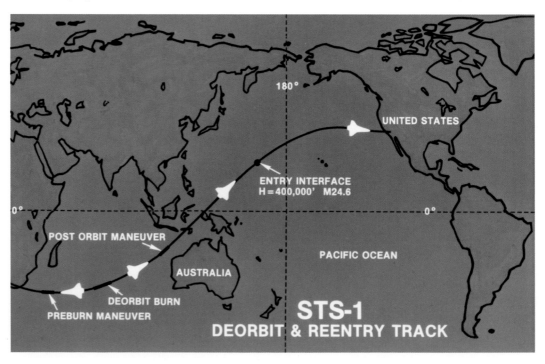

STS-1 DEORBIT & REENTRY TRACK

RIGHT Flying around the heading alignment cylinder to convert the entry path into the landing approach pattern, *Columbia* touched down 54hr 20min 53sec after launch. *(NASA)*

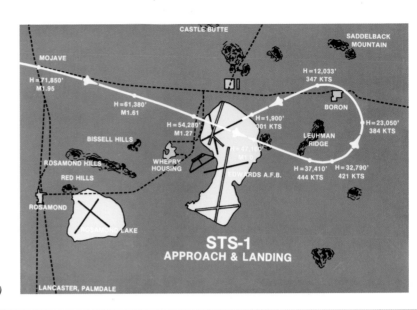

destroyed by the fiery heat of re-entry. Exactly two minutes after the three main engines shut down the two Orbital Manoeuvring System (OMS) engines, using propellant housed in tanks within the two blisters either side of the tail, fired for 86 seconds to put *Columbia* in a

BELOW Touchdown came on Runway 23, at a speed of 207mph, rolling to a stop in 8,993ft. *(NASA)*

MAXIMUM CAPACITY 4 PERSONS OR 1000 LBS.

Columbia

safe orbit. Another firing of the OMS engines for 75 seconds, 44 minutes after launch, placed *Columbia* in a 153 x 154 mile orbit.

After two days of orbital tests *Columbia* turned to place its aft end in the direction of travel, pitched the nose down slightly so that the engines at the rear were pointing upward, and fired the two OMS engines in a retro-burn that slowed the Shuttle and gradually brought it down toward the earth. Turned around now, and pitched up slightly and began to enter the atmosphere at a height of 76 miles still some 5,000 miles up-range of its landing strip at Edwards Air Force Base toward which it would fly, gradually decelerating

from nearly 18,000mph to a touchdown speed of 207mph. With a weight of just under 200 tons (197,472lb) *Columbia* rolled to a stop in 8,893ft and was delivered back to the Kennedy Space Center atop the Boeing SCA on 28 April.

Building on success

Work to build the second space flight Orbiter was commissioned in a contract dated 5 January 1979. Beginning on 28 January 1979, STA-099 was upgraded for orbital flight, re-designated OV-099 and named *Challenger*, after the British naval research vessel that sailed the Atlantic and Pacific oceans in the 1870s.

On 29 January 1979, NASA issued a contract to Rockwell International for two more Orbiters, OV-103 and OV-104. Work on OV-103 started on August 29 that year. Named *Discovery*, after the ship used by Captain Cook in the 1770s that led to the discovery of the Hawaiian Islands, it benefited from lessons learned building *Enterprise*, *Challenger* and *Columbia*. Whereas *Columbia* had a dry weight of 178,289lb, *Discovery* weighed 171,419lb. The fourth Orbiter, OV-104, was named *Atlantis*, commemorating the primary research ship of the Woods Hole Oceanographic Institute from 1930 to 1966. Assembly began on 3 March 1980.

A fifth Orbiter was planned, with assembly beginning on 15 February 1982, but work was stopped in April 1983 when NASA decided to save money and complete work that would only constitute a set of spares and not a complete flight rated vehicle. Then, when *Challenger* was destroyed on 28 January 1986, Congress approved a fifth space flight Orbiter, and work to build OV-105 recommenced on 28 September 1987. Named *Endeavour*, after another of Captain Cook's vessels, it joined the remaining three surviving Orbiters with rollout on 25 April 1991, and made its first flight into space on 7 May 1992.

Endeavour was the only Orbiter fitted out for remaining up to 28 days in space but the equipment to do so was subsequently removed to save weight. But it did carry several innovations, including a tail parachute to assist with rollout braking after touchdown. It also pioneered new technology that would eventually be retrofitted to the other Orbiters. Various efforts to expand the fleet were made, most noticeably in 1988 when NASA considered a second new Orbiter in addition to *Endeavour*, thereby maximising the reopened assembly line and providing a space station rescue vehicle. That came to nothing, and neither did another attempt a year later to assemble a set of spares, much as had been done with the elements built into *Endeavour*. Sadly, once again the Shuttle fleet reverted to three operational Orbiters with the loss of *Columbia* on 1 February 2003.

ABOVE Taken in orbit, this image shows tiles missing from the right aft Orbital Manoeuvring System pod, some of the 16 torn away and 148 damaged during launch. *(NASA)*

CENTRE Streaking caused by the heat of re-entry is plainly visible on the main landing gear door. *(NASA)*

LEFT Forever symbolising the world's first reusable manned space vehicle, a pinnacle of aeronautical achievement, the mission badge of STS-1, flown by Young and Crippen. *(NASA)*

Chapter Three

Anatomy of the Shuttle

The Shuttle Orbiter is a reusable vehicle intended to carry astronauts and cargo to and from space. It is about the size of a DC-9 airliner and is designed to survive the rigours of launch and landing, including vibration, high acoustic levels from the rocket engines, high levels of acceleration and various heat loads on different parts of the structure. The layout is dominated by just two requirements – to carry a design payload of up to 65,000lb to orbit, and to fly back down through the atmosphere like an aircraft, landing like a glider so that it can be used again.

LEFT The pressure vessel, comprising the habitable flight deck and mid-deck areas, is lowered into the lower half of the forward fuselage. *(North American)*

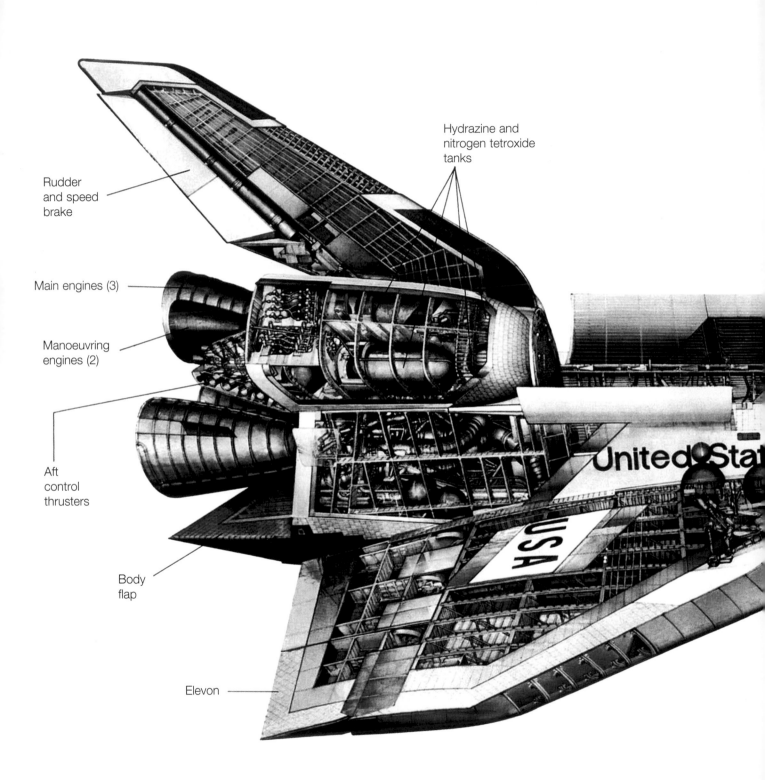

Rudder
and speed
brake

Hydrazine and
nitrogen tetroxide
tanks

Main engines (3)

Manoeuvring
engines (2)

Aft
control
thrusters

Body
flap

United Stat

USA

Elevon

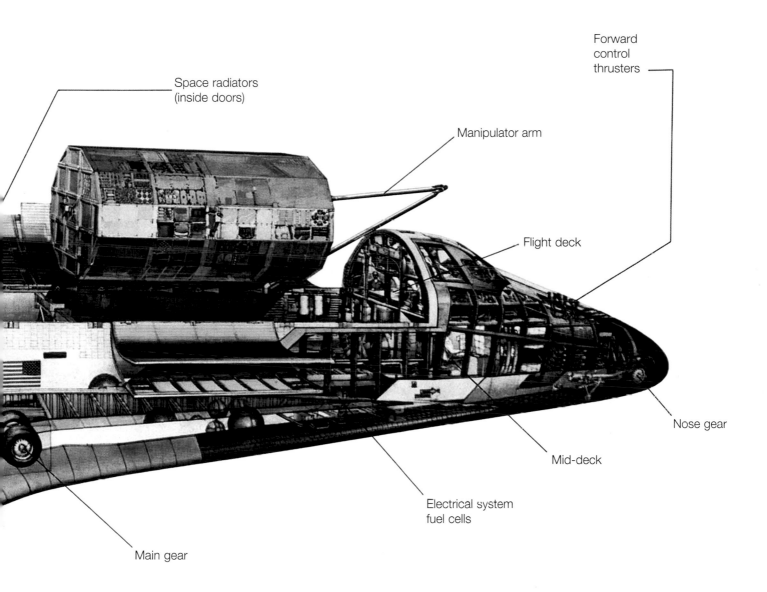

Space radiators
(inside doors)

Manipulator arm

Forward
control
thrusters

Flight deck

Nose gear

Mid-deck

Electrical system
fuel cells

Main gear

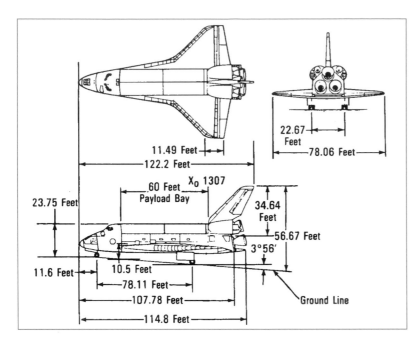

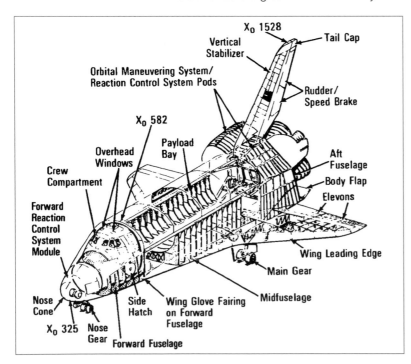

ABOVE The critical dimensions of the Orbiter were shaped by the requirement to provide a cargo bay 60ft long by 15ft in diameter. *(NASA)*

Because of these requirements the Shuttle is shaped to look like an aircraft but to operate as a spacecraft. The structure of the Shuttle Orbiter comprises nine separate sections, or elements: the forward fuselage, the forward reaction control system module, the mid-fuselage, the payload bay doors, the aft fuselage, the vertical tail, the two orbital manoeuvring system/reaction control modules and the wing.

The demands are greater than is usually the case with a conventional aircraft because the stresses imposed upon the structure are unique to the Shuttle. Because of this, the design team at North American Aviation had no precedents on which to base their prototype. It was the first of its kind, without the advantage of any previous learning curve, and one of a kind without parallel.

Very few aircraft designed for operational use break completely new ground in their operating environments. Two may be considered as such: the Mach 3 Lockheed SR-71 spy plane and the 1,400mph Concorde, the world's first commercially viable supersonic airliner. But the Shuttle would follow its own development path. At first it was an experimental vehicle designed to be adapted later to operational requirements, which included carrying satellites into space. It would also be called upon to lift large modules into orbit for a space station, and carry a wide range of satellites and spacecraft to be deployed in different trajectories, some of which would be sent to the outer regions of the solar system by the rocket motors attached to them.

Forward fuselage

This consists of upper and lower sections divided horizontally, which fit like a clamshell over the pressurised crew compartment where the astronauts live and work when they are not space walking or transferring to another spacecraft. The forward fuselage is fabricated from 2024-T81 aluminium alloy with skin-stringer panels, frames and bulkheads. The stringers are located 35in apart while the vertical frames are 30–36in apart, riveted to the stringer panels.

The pressurised crew compartment is attached to the forward fuselage at four locations. It is of welded construction to achieve an air-tight pressure vessel capable of providing a shirt-sleeve environment and of sustaining the crew with an atmosphere of oxygen and nitrogen at sea-level pressure (14.7lb/sq in).

The crew compartment has three levels.

LEFT The distribution of spacecraft systems is accommodated within a generally simple structural design, using standard aircraft manufacturing practices. *(NASA)*

ABOVE Mated to the mid-fuselage section, the forward fuselage accommodates the pressure vessel that the crew will inhabit, the forward reaction control system and the nose landing gear. *(North American)*

There is only one way in or out of the Orbiter on the ground, through the 40in diameter circular side hatch which, with the Orbiter on its landing gear, opens downwards or, with the Orbiter on the launch pad, to the side. It can also be used to escape from the Orbiter if it is unable to land after re-entering the atmosphere. The mid-deck area is accessed directly when the vehicle is on the ground, with the flight deck above and the equipment bay below. In weightlessness, access to the flight and mid-deck areas is a matter of simply floating through one of two hatches, each 26in x 28in. The pressurised crew compartment is 17½ft high, 16½ft long and the forward cylindrical nose section is 10.6ft in diameter. It also has provision for an airlock that allows astronauts to leave the crew compartment and move into the unpressurised cargo bay, which forms the main section of the mid-fuselage assembly.

There are 11 main windows in the crew compartment: 6 wrapped around the forward area of the flight deck, 2 in the aft bulkhead, which faces directly into the payload bay, 2 in the roof of the flight deck and 1 in the side

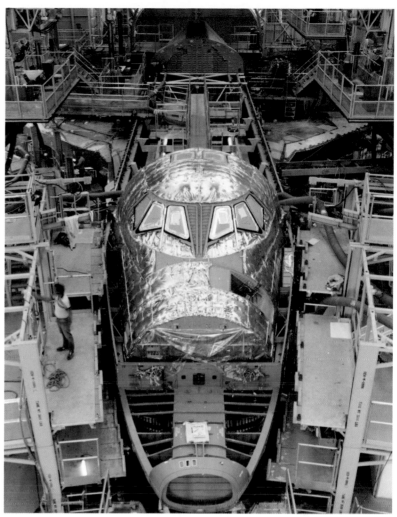

ABOVE Insulated to minimise the extremes in temperature that the environmental control system must handle, the pressure vessel rests inside the nose shell of the forward fuselage.
(North American)

LEFT Encapsulating the pressure vessel, the upper shell of the forward fuselage has window elements that mate with the window frames on the pressure vessel.
(North American)

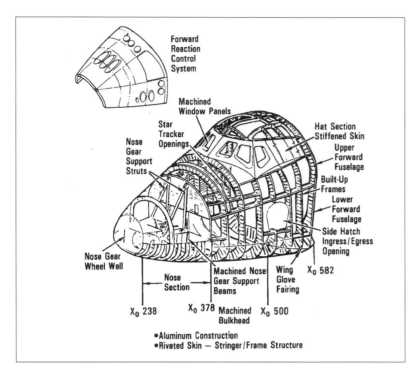

Forward Reaction Control System

Machined Window Panels

Star Tracker Openings

Nose Gear Support Struts

Hat Section Stiffened Skin

Upper Forward Fuselage

Built-Up Frames

Lower Forward Fuselage

Side Hatch Ingress/Egress Opening

Nose Gear Wheel Well

Nose Section

Machined Nose Gear Support Beams

Wing Glove Fairing

X_0 582

X_0 238

X_0 378

Machined Bulkhead

X_0 500

• Aluminum Construction
• Riveted Skin — Stringer/Frame Structure

ABOVE The forward reaction control system module is detachable for servicing, and its removal enables access to the lower nose section and upper landing gear bay. *(NASA)*

hatch on the left side of the crew compartment in the mid-deck area. The forward-facing windows are used by the two pilots for entry and landing as well as some on-orbit operations. The two rear-facing and upper-facing windows are used for rendezvous and docking manoeuvres and for observing activity in the payload bay.

The six forward windows are the thickest ever assembled with optical quality and comprise three separate panes: the innermost

for withstanding crew compartment pressure, the middle one providing an optically transparent thermal shock layer, and the outer pane providing both thermal and impact protection. Both inner and outer panes are each 0.6in thick. The inner and middle panes are attached to the crew compartment while the outer pane is attached to the upper section of the forward fuselage.

The total interior volume of the crew compartment is 2,325cu ft and the atmosphere, maintained at 14.7lb/sq in, is a constant 80/20 mix of nitrogen and oxygen. Usually, four seats are provided on the upper flight deck with a further three seats on the mid-deck area. Although additional seats could be installed for emergencies or for exceptional needs, the Shuttle usually flies with a complement of seven astronauts. The two pilots' seats (the left seat being the commander's position) are occupied for all launch, re-entry and major propulsive burns in orbit. The other seats are for mission specialists – astronauts who are not necessarily selected for their piloting skills, but who are there to conduct mission operations and sundry scientific tasks, as well as to assist with moving payloads in or out of the cargo bay and to perform space walks (called EVA or extra-vehicular activity). Mission specialist seats are stowed during orbital operations and re-installed for re-entry and landing.

According to mission requirements, bunks can be installed in the mid-deck area as well

RIGHT Technicians carry by hand into the pressure vessel, the several miles of Orbiter nose section wiring. *(North American)*

as a galley for food preparation. The waste management facility (toilet) is installed in the mid-deck, too, and this area provides 140cu ft of stowage area with modular lockers for astronaut gear, personal hygiene equipment and for experiments – in all, 42 identical boxes, each 11in x 18in x 21in.

Below the mid-deck area is the equipment bay. It is here that the astronauts can gain access to waste management equipment, air revitalisation systems, pumps, fans, lithium hydroxide canisters for removing carbon dioxide breathed out by the crew, together with an additional five spaces for extra crew equipment stowage.

The mid-deck area also serves to house the cylindrical airlock, with an interior diameter of 5ft 3in and a length of 6ft 11in, and two 40in diameter circular openings and pressure-tight hatches. One hatch is on the front facing inside the mid-deck, the other on the opposite side of the airlock and is attached direct to the aft bulkhead which, upon opening, allows access into the payload bay. The airlock can also be installed on the inside of the payload bay, attached to a tunnel adapter leading to a pressurised research module such as Spacelab or Spacehab, where the astronauts can work in a shirt-sleeve environment on scientific experiments carried up from the ground and installed in racks. The airlock is big enough to contain two fully suited crewmembers simultaneously.

The forward fuselage also supports the reaction control system module (RCSM), which carries the nose thrusters for attitude control in space. This section is removable for servicing, replenishing the propellant (fuel and oxidiser) tanks and attending to the plumbing. The RCSM is removed and serviced in the Orbiter Processing Facility (OPF) where the vehicle is turned around after each flight and made ready for the next launch. The forward fuselage also contains the forward landing gear.

Escape from the crew compartment is possible during descent when the Orbiter is off target and likely either to ditch or to crash-land without reaching a runway, but only if it is in a controlled glide. Because of the shape of the Orbiter and its large delta wing, an astronaut leaping from the side hatch would in all probability strike the leading edge of the wing itself. To throw the astronaut beyond the wing, an escape pole can be quickly fitted to the inside of the Orbiter mid-deck, extended to its full length of 10½ft and projected through the open hatch. Wearing a partial pressure suit and with a parachute, the astronaut would place a looped lanyard over the pole and leap from the side hatch. Instead of being thrust back against the wing or the fuselage by the slipstream, the lanyard and attached astronaut would slide down the pole and be catapulted in a slingshot manoeuvre away from the Orbiter.

Commitment to the emergency escape method would be made with the Orbiter descending through 60,000ft; when the Orbiter reaches 30,000ft the speed has reduced to 230mph. At about 25,000ft a crewmember nominated as jump master prepares the equipment, and the flight control system on the Orbiter maintains the angle of attack at 15°. With the escape pole inclined downwards from the side hatch it would take only 90 seconds for all seven crewmembers to get free, the last at an altitude of about 10,000ft. The Orbiter would crash, but the crew would have escaped. Two additional emergency escape procedures cover situations on the ground after landing, via an escape slide from the side hatch and out of an emergency escape hatch in the roof of the flight deck.

BELOW Like furniture movers unable to find a big enough door, the technicians pass the wiring loom through the starboard side window in the upper fuselage shell.
(North American)

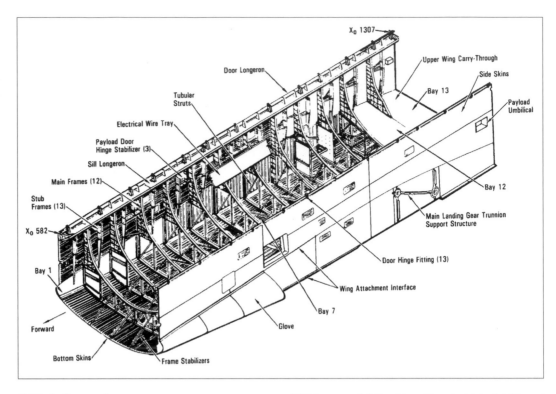

Labels on diagram:
X₀ 1307
Upper Wing Carry-Through
Side Skins
Bay 13
Payload Umbilical
Door Longeron
Tubular Struts
Electrical Wire Tray
Payload Door Hinge Stabilizer (3)
Sill Longeron
Main Frames (12)
Stub Frames (13)
X₀ 582
Bay 1
Forward
Bottom Skins
Frame Stabilizers
Glove
Bay 7
Wing Attachment Interface
Door Hinge Fitting (13)
Main Landing Gear Trunnion Support Structure
Bay 12

Mid-fuselage

This is the structural backbone of the Orbiter, incorporating the cargo bay and its doors, and provides support for the forward fuselage, the aft fuselage and the wings. Built by General Dynamics in San Diego, California, it is about

60ft long and 17ft wide with a height of 13ft and it weighs about 13,500lb. It is assembled from twelve main, vertical, frame assemblies, each with special weight-saving boron/aluminium trusses for strength, with reinforced skin and longerons. The two top edges of the mid-fuselage are especially strong. In addition to supporting the sills for the payload bay doors, they also take bending loads for the entire Orbiter and it is from these and the longerons that the payloads are 'hung'. Unique at the time the Orbiters were first manufactured, skins for the mid-fuselage were machined integrally by numerical control.

The floor of the mid-fuselage consists of the wing carry-through box and the wings themselves are attached to the outer surface of this section. The first few Shuttle flights provided important measurements about stresses and temperatures that were higher than expected when the designers put together the mid-fuselage. To strengthen this area, engineers attached torsional straps to tie together all the stringers thereby, in effect, creating what amounts to a box-section. Early Orbiters were retro-fitted with vulcanised silicone rubber inserts to absorb heat and distribute it more uniformly across the lower section of the mid-fuselage structure.

The payload bay (or cargo bay as it is sometimes known) is capable of handling equipment up to 60ft long and 15ft in diameter and is covered by two doors, left and right of the centreline. Each door is assembled from five sections connected by circumferential expansion joints and connected to the mid-fuselage sill by 13 hinges, of which 8 are 'floating' to allow expansion and contraction of the mid-fuselage section due to mechanical stresses and to the wide variations in temperature the Orbiter experiences in space. Five hinges are fixed.

The payload bay is a relatively air-tight structure, achieved by means of a seal that runs right around the outer edge of each door. The payload bay is not pressurised, but the seals prevent leakage of heat from the top of the Orbiter to the interior. Because the Orbiter re-enters the atmosphere nose-up, the payload bay doors experience relatively low temperatures, shielded as they are from the build-up of heat on the main undersurfaces of the Shuttle.

Each door is 60ft long by 10ft across the radius, locked down onto the rest of the structure by 16 latches along the centreline and 8 latches at each end, the forward fuselage and the aft fuselage. The doors are manufactured

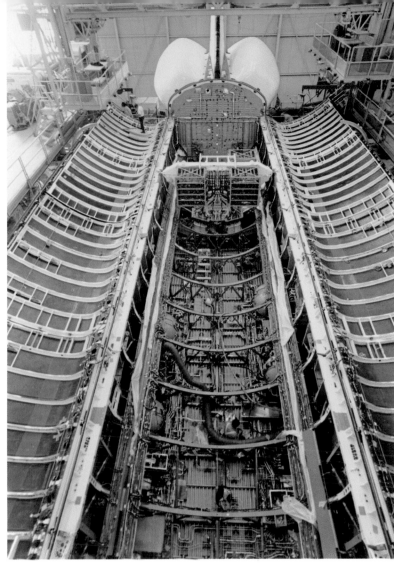

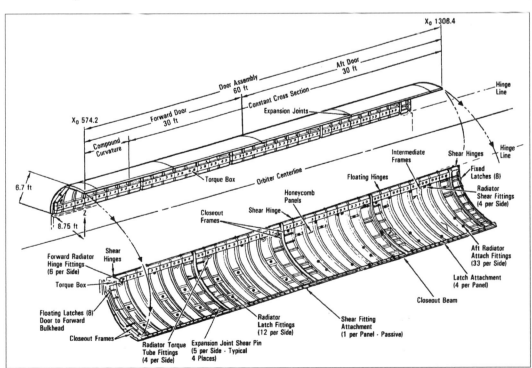

ABOVE The payload bay is half the total length of the Orbiter, and is designed to cope with the flexing loads it will experience in space from extremes of temperature.
(North American)

LEFT The cargo bay doors are not designed to carry loads from the centre fuselage, but they do carry radiator panels to help maintain a balanced thermal environment.
(NASA)

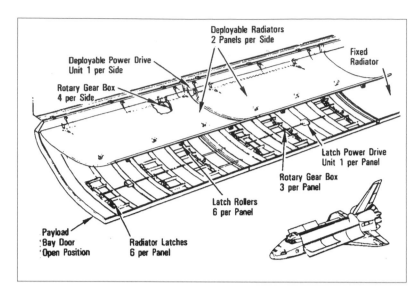

Deployable Radiators
2 Panels per Side

Deployable Power Drive
Unit 1 per Side

Rotary Gear Box
4 per Side

Fixed Radiator

Latch Power Drive
Unit 1 per Panel

Rotary Gear Box
3 per Panel

Latch Rollers
6 per Panel

Payload Bay Door
Open Position

Radiator Latches
6 per Panel

ABOVE The payload bay doors can be manually closed if the hydraulics fail, and the radiator panels can be hinged upward to provide a cooling surface on both sides. *(NASA)*

from a composite material of graphite-epoxy and Nomex, saving about 23 per cent of the weight of a similar structure made from aluminium. Because the right-hand door (as viewed from the top looking forward) carries the latching mechanism, it weighs 2,535lb, compared with 2,375lb for the left-hand door. The doors open through an angle of 175.5°.

The interior surfaces of the payload bay doors each support four radiator panels running down the entire length of the bay, each panel measuring 15ft long and 10ft across the curve.

They are there to control the amount of heat removed from the interior of the Orbiter and ejected into space to reduce overheating from the spacecraft's systems or from the sun's energy. The two forward panels are hinged so as to tilt upwards by about 35° from the payload bay doors when they are opened after reaching orbit, thereby allowing heat to be ejected from both sides of the radiator. The two rear panels are fixed, but extra panels can be carried for some missions.

The radiators were manufactured by the LTV Corporation (now part of Lockheed Martin), and NASA upgraded them after delivery to prevent damage from micrometeoroids. They comprise a series of tubes, each of the four deployable panels carrying 68, with each of the 4 aft (fixed) panels supporting 26 tubes, all with a diameter of about 0.1in. The tubes are connected to the Freon coolant loops, separate systems on respective door panels, which control the temperatures inside the Orbiter. The modifications consisted of thin metal strips placed above each tube to prevent an impact from space debris creating a puncture and threatening the mission.

Aft-fuselage

This section comprises a structural mount for the vertical tail, the hinged body flap and a so-called thrust structure, containing the three main rocket engines and the plumbing necessary to bring fuel and oxidiser from the External Tank. It also supports the two removable orbital manoeuvring system/reaction control system (OMS/RCS) pods. These pods carry propellant tanks, plumbing and rocket motors for manoeuvring in space and for keeping the Orbiter at the correct attitude in orbit and during the early phases of re-entry, where the air pressure is too low for the aero-surfaces to control the vehicle like an aircraft. Manufactured primarily out of graphite epoxy composite material and aluminium, each pod

LEFT The aft-fuselage section is dominated by the thrust structure, designed to transmit loads from the three main engines. The orifices for the main engines are clearly visible. *(North American)*

RIGHT The aft-fuselage section is manoeuvred into position, ready to attach to the mid-section. The body flap will be attached at the lower rear edge. *(North American)*

measures 21.8ft in length and 11.4ft wide at its after end, and approximately 8.4ft wide at the forward end.

Like the mid-fuselage, the aft-fuselage is also a load-bearing structure, transmitting forces from the main engines up through the Orbiter and across the rear face of the wing carry-through structure, which is part of the mid-fuselage. While the mid-fuselage acts as a strong-back for the Orbiter, the aft-fuselage serves to transmit loads from the three main engines to the Orbiter and the External Tank. Comprising an outer shell, thrust structure and internal secondary structure, the aft-fuselage is about 18ft long, 22ft wide and 20ft high. It serves as an interface with the main wing spar and provides the aft closeout bulkhead at the forward end with the mid-fuselage and payload bay, comprising machined and beaded sheet metal aluminium segments. The upper part of the bulkhead attaches to the front spar of the vertical tail.

The internal thrust structure serves as a support for the three main engines (SSME) with a load reaction truss, engine fittings and the actuator support structures. The aft fuselage also supports the SSME low-pressure turbo pumps and propellant lines and attachment points for connecting the Orbiter to the External Tank. The internal thrust structure is primarily 28 machined, diffusion-bonded truss members, where titanium strips are bonded under heat and pressure which, over time, fuses them into a single, hollow mass that is much lighter and much stronger than a machined part.

The outer shell of the thrust structure is formed from integrally machined aluminium. Exposed areas are covered with thermal protection materials to help insulate the structure from the heat of re-entry. A secondary structure fabricated from aluminium supports a variety of brackets, webs, machined fittings and avionics bays, which are shock-mounted to the structure itself.

The body flap is a 21ft-wide aluminium structure where it attaches to the aft-fuselage, and 18¼ft wide at the trailing edge. It is 7.24ft

long and can be pivoted 15.7° up and 26.55° down so that it can serve as a pitch trim for the Orbiter during its descent through the atmosphere. The body flap also serves to shield the three SSMEs from the heat of re-entry.

Vertical tail

This consists of a fin 26ft 4in tall incorporating a split rudder/speed brake, the fixed portion of the fin being built up from aluminium ribs and spars and attached to the

BELOW Fairchild built the vertical tail assembly, arguably the only functional element that has no role other than that required for directional control and stability after re-entry. *(North American)*

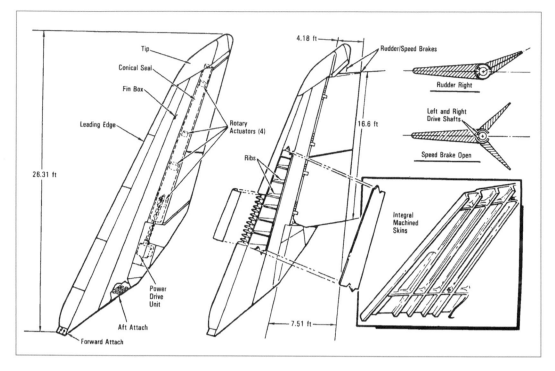

RIGHT The tail itself brought unique challenges, the extremes of temperature imposing thermal stresses on what is mostly a moving surface. *(NASA)*

BELOW The thickness/chord ratio of the wing is clearly evident in this pre-assembly view. *(North American)*

upper structural surface of the aft-fuselage. The rudder is 16.6ft tall and 7½ft wide at the base, and is of similar construction to the fin but of two separate halves co-located at the hinge line. As a rudder the two closed surfaces can move 27° either side of the centreline or, when operating as a speed brake, the drive shafts turn in opposite directions to spread the two halves of the rudder to a maximum 49.3° each, presenting a spread of 98.6°.

Wing

The Orbiter's wing is the aerodynamic lifting surface that provides conventional lift and control for the vehicle when it is within the earth's atmosphere. Each wing has a glove, an intermediate section, a torque box, a forward spar to which is attached the leading edge thermal protection, elevon surfaces along the trailing edge and a main landing gear well. The wing itself is built up in a multi-rib and spar arrangement, with stiffened stringers supporting the exterior skin. Each wing has a length of about 60ft and a thickness of 5ft and the forward wing box aerodynamically blends the leading edge into the mid-fuselage wing glove, or fillet. This is made up from aluminium ribs, tubes and tubular struts.

The wing contains four major spars, each constructed of corrugated aluminium to minimise thermal loads. The forward spar is used to attach the curved Reinforced Carbon-Carbon (RCC) heat shield and to form the rounded profile of the wing leading edge. The rear spar is the attachment surface for the trailing edge elevon hinges. The two-piece elevons are of conventional aluminium rib and

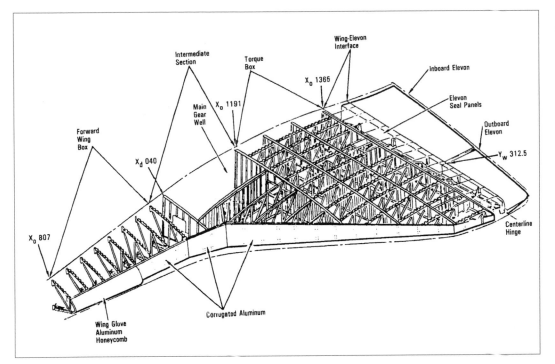

LEFT The broad torque box provides the wing carry-through junction at the attachment to the centre fuselage, the aluminium facing panels on the leading edge providing a mount for the carbon-carbon heat shielding. *(NASA)*

beam construction and are divided into two segments per wing, each segment supported by three hinges. Attached to the flight control system hydraulic actuators, each elevon travels a maximum of 40° up and 25° down.

Each outboard elevon is just over 12ft long, 3¾ft wide at the outermost edge and 6ft at the inboard edge where it lies adjacent to the

LEFT Attachment location points for the 22 carbon-carbon thermal protection panels around the wing leading edge. *(North American)*

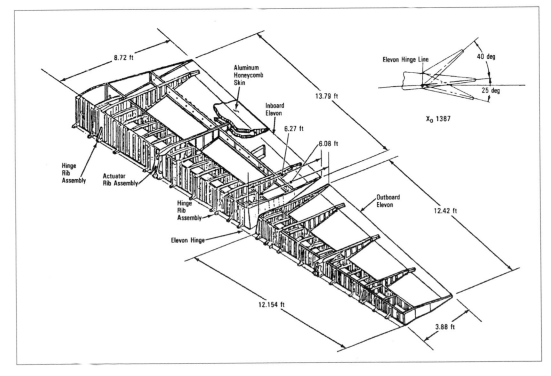

LEFT: The trailing edge supports the inboard and outboard elevons, which combine the traditional functions of ailerons, (for roll), and elevators, (for pitch). *(NASA)*

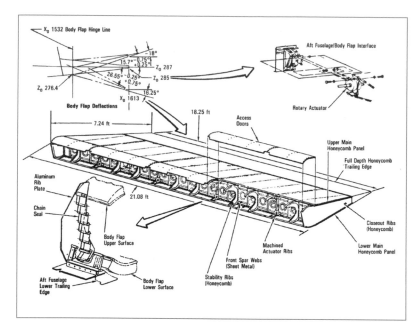

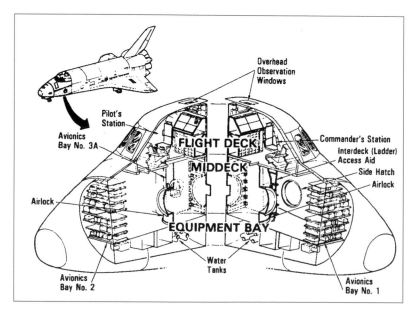

inner elevon, 13.8ft long and 8.7ft wide at the inboard edge. The main landing gear doors are 5ft wide and 12.6ft long and are located in the wing intermediate section. During the weight reduction programme prior to the assembly of wings for OV-103 (*Discovery*) and OV-104 (*Atlantis*), certain areas of the structure were redesigned as a result of loads measured in flight by previous Orbiters and found to be greater than predicted. Doublers and stiffeners were applied to OV-102 (*Columbia*) and OV-099 (*Challenger*) to maintain safety levels.

Crew equipment

Life aboard the Shuttle is very different from that experienced by astronauts on the first generations of spacecraft, which were small capsules with little or no room and primitive conditions. Eating, sleeping, personal hygiene and work were conducted in a two-man vehicle where each had about the amount of room available in the front of a small car. That was all the space afforded to astronauts in the Gemini programme, where ten pairs of astronauts rehearsed space techniques prior to Apollo during 1965 and 1966. On one of those flights – Gemini 7 – two astronauts spent 14 days in the equivalent space afforded by the front seats of a family car!

Food came in tubes, toilet was conducted with nappies and stick-on bags, pure oxygen puffed up the face and produced strange odours, and all the while the whirring of pumps, the occasional clank of a valve opening or

RIGHT Located on the starboard side of the mid-deck area, sleep stations have gone through several incarnations, depending upon the number of crew members and the mission. *(NASA)*

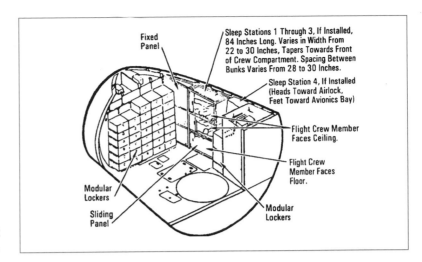

closing and the overriding stench accumulating day after day with little or no result from the 'odour removal' system. That was life aboard spacecraft that carried men around the earth on flights of up to 14 days and later, just when the Shuttle was conceived, to the surface of the moon and back. But the Shuttle was so different it was akin to being on the flight deck of a cargo plane, with another deck below in which to gather together for a meal, use wash facilities and lead the semblance of a normal life.

It had to be that way because the Orbiter would be home to between six and eight astronauts working in the mid-deck area on a variety of scientific experiments, or through the airlock module into a research laboratory bolted into the spacious cargo bay. The demand on Orbiter crew equipment was much greater than on previous spacecraft and included personal hygiene items, bedrolls, suits, food, drinks, water and cabin refreshment. All in a vehicle designed with a standard two-gas atmosphere at comfortable room temperatures. Indistinguishable from the atmosphere breathed on earth.

In Apollo spacecraft three men could live for up to 12 days, a capability said to be 36 man-days. With a standard capability of supporting

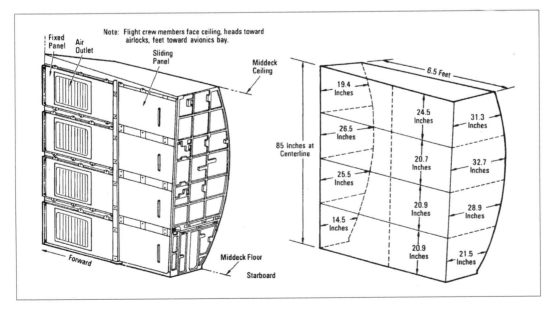

ABOVE The general-purpose computers are located in the forward avionics bay, and contain the core guidance and navigation systems. *(North American)*

LEFT Aft of the forward avionics bay, the modularised sleep compartments give scant privacy in what can be a very crowded space for up to seven people. *(NASA)*

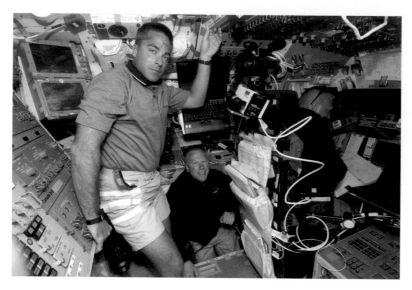

ABOVE Access from the flight deck to the mid-deck area is via a hatch, through which astronauts float. Note the flight reference documents on the back of the commander's seat. *(NASA)*

ABOVE Throughout the flight, the commander and the pilot use laptops to practice landing techniques, computer simulations honing flying skills. *(NASA)*

eight astronauts in space for up to 18 days, the Shuttle was designed to support 144 man-days. Additional life-extension kits were available for even longer missions, but these were not standard and were never used to their full capacity. Nevertheless, it placed huge demands on the engineering design of the interior, adopting aesthetic colours and provision for personal mementoes and decorations from home.

Such things were important and during the early design phase several consultants provided valuable advice on how best to accommodate people far from home with no easy way back. In 1973 and 1974, NASA's Skylab space station used redundant Apollo hardware to provide a large, spacious work station and living quarters for teams of three astronauts on three separate missions lasting 28, 59 and 84 days, respectively.

BELOW The port rear flight deck area showing the upper window (top) and one of the two windows facing directly back into the cargo bay. *(NASA)*

BELOW Astronaut Clayton Anderson aboard *Atlantis* prepares to film views literally out of this world. Note the hand controller at left. *(NASA)*

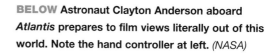

Design of equipment and preparation for Skylab ran parallel to development of early Shuttle configurations (*see Chapters 1 and 2*).

Because Skylab would support astronauts for several months in space, considerations of interior design and colour were important. NASA consulted the style designer Raymond Lowey, famous for the Coke bottle, several designs of Studebaker car and a host of iconic American brands from the 1950s and 1960s. He showed how subtle changes with no impact on engineering criteria could make life more relaxed and 'homely'.

The Shuttle would be a relatively short-duration vehicle but these considerations were fed in to the overall selection of style and form for all the many domestic activities essential to running a tidy ship. Although a legacy from the capsule days of manned flight, the term 'housekeeping' developed a unique meaning all of its own for space-bound astronauts living in a small group in a two-storey space plane. For that reason the Shuttle became the first integrated launch and re-entry vehicle to be work place, home and assembly station for astronauts in orbit. The selection of food type, dietary intake and relaxation cycles were equally important to engineering considerations.

The food and dining system aboard the Shuttle was based around a range of foods in

LEFT As old as humankind, the practice of gathering together to eat, carried into space by the Shuttle, is enjoyed by the multinational crew of the International Space Station in their more spacious modules. *(NASA)*

LEFT In the mid-deck area, with storage lockers to the right and the ingress/ egress hatch at lower left, astronaut and future director of the Kennedy Space Center, Robert Cabana, floats aboard *Discovery* on STS-41. *(NASA)*

BELOW In weightlessness, astronauts' arms must be restrained for fear of taking on a zombie-like appearance – hanging seemingly lifeless in front of the torso! *(NASA)*

BELOW Astronauts Clark, Husband and Chawla in mid-deck sleep bunks during their flight aboard *Columbia*, an image recorded prior to the loss of *Columbia* on 1 February, 2003. *(NASA, with permission of the families)*

ABOVE Looking towards the forward, starboard quarter of the flight deck aboard *Endeavour*, during the STS-130 mission in February 2010, with the pilot's seat in the centre. *(NASA)*

specific storage categories. For this purpose the mid-deck area was to be the dining room where food was also stored and prepared around a standard three meals per day.

Thermal protection system

R eturning from space creates heat through friction with the earth's atmosphere. While early science fiction writers, and designers of

early hypersonic experimental aircraft, thought the best way to survive this was to design a needle-pointed object that could rapidly swoop down to a landing, reality proved otherwise. A pointed object attracts shock waves that attach directly to it. Kinetic energy produced by atmospheric friction is converted into heat, which is so extreme that it is beyond the capability of most materials to survive.

In the 1950s an aerodynamicist named Alfred Eggers found a solution. He showed that an object shaped like a flat-iron is best for surviving re-entry. By entering the atmosphere bottom first it would create a shock wave in front of the blunt base, detached from the object and preceding it as it descended. The shock wave would contain most of the heat that would otherwise attach itself to the surface of the spacecraft, which would instead only get the radiated heat from the shock wave. Nevertheless, temperatures at the surface would reach almost 3,000°F, but within the survivable range of known materials. This is why all early spacecraft entered base-first to create a stand-off shock wave and separate the primary source of heat.

All US-manned spacecraft prior to the Shuttle used ablative heat shield materials,

RIGHT The black thermal protection tiles are individually shaped, numbered and allocated a specific location across the broad underbody and undersurface of the wing. *(NASA)*

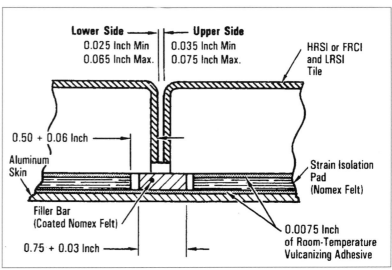

Diagram labels:
Lower Side
0.025 Inch Min
0.065 Inch Max.

Upper Side
0.035 Inch Min
0.075 Inch Max.

HRSI or FRCI and LRSI Tile

0.50 + 0.06 Inch

Aluminum Skin

Strain Isolation Pad (Nomex Felt)

Filler Bar (Coated Nomex Felt)

0.75 + 0.03 Inch

0.0075 Inch of Room-Temperature Vulcanizing Adhesive

designed so that some of the surface would burn away and carry most of the heat with it. Re-entering the earth's atmosphere at around 25,000mph on return from the Moon, the Apollo Command Module (CM) had to encounter a particularly severe environment. To protect it the entire outer surface of the CM's heat shield was a stainless steel honeycomb, the cells filled with an epoxy to char away as it re-entered the atmosphere. Because all previous manned spacecraft were used only once, the ablation process was acceptable, but the Shuttle was very different and would be used repeatedly with a stipulated requirement for minimal servicing between flights. The heat shield would have to survive intact and remain usable many times over.

Instead of ablatives, NASA chose to use so-called heat-sink materials to protect the aluminium structure of the Orbiter, which is limited to a maximum temperature of 350°F. From the perspective of protecting the Orbiter from high temperatures, the best choice for the Shuttle would be Reinforced Carbon-Carbon (RCC). Used in the re-entry nose cones of ballistic missiles, it is a composite comprising carbon fibre reinforced in a matrix of graphite and has found widespread application. Developed for motor racing by the Brabham team in the 1970s, it is now a standard material used in the disc brakes of Formula 1 cars.

RCC is ideally suited to high thermal shock where a low coefficient of expansion is called for – in other words, if rigid it works well but if it is required to flex it shatters. It is also relatively

ABOVE LEFT The edges of doors and panels are particularly sensitive to heat creep. Some, like the forward landing gear doors, occupying a considerable surface area. *(NASA)*

ABOVE Because the structural frame of the Orbiter flexes and bends according to thermal loads during flight and in space, the tiles have miniscule gaps between them to allow expansion and contraction without cracking. *(NASA)*

BELOW A large percentage of the thermal protection tiles are retained for many flights, but some are replaced after each landing, and every one is meticulously examined for signs of damage, stress or abrasion. *(NASA)*

brittle and susceptible to shock. Although it would have been desirable to cover the exterior of the Orbiter in RCC, with a density of 124lb/cu ft, it is prohibitively heavy. Moreover, because it is inflexible, the thermal expansion and contraction of the Orbiter structure would cause it to crack. In the vacuum of space where there is no conductivity, the Shuttle experiences wide changes in temperature from −180°F to +180°F and the structure flexes by small amounts as it expands and contracts.

Recognising that heat and temperature are two separate conditions, RCC was essential for the hottest places with the highest temperatures, such as the nose of the forward fuselage and the leading edge of the wings. The nose of the Orbiter would not flex and a rounded section of RCC was possible for that. But the wings would flex so the RCC was applied to the leading edge as 22 separate U-shaped sections on each wing, with floating joints to permit a degree of wing flexing. RCC could withstand temperatures as low as −250°F and would protect the hottest regions of the Orbiter from the highest temperatures on re-entry, expected to reach almost 3,000°F.

For protecting the Orbiter from temperatures below 2,300°F, two different types of individual insulation tiles were applied. For the hottest regions, where temperatures can reach above 1,200°F, High temperature Reusable Surface Insulation (HRSI) tiles are fitted. For regions where temperatures reached a maximum 1,200°F, Low temperature Reusable Surface Insulation (LRSI) are used. The black-coated HRSI comprises individual brick-like tiles with a thickness of 1–5in, depth depending on location but generally thicker forward and thinner to the rear of the wing and fuselage undersurfaces. Each HRSI tile is about 6in x 6in square and is manufactured from 99.8 per cent pure silica, derived from high quality sand, with a tile density of 10 per cent fibres and 90 per cent air. These HRSI tiles have a density of 22lb/cu ft.

Another form of insulation serving the same temperature band is known as the Fibrous Refractory Composite Insulation (FRCI) tile, developed at NASA's Ames Research Center in California. It has much lower mass than the standard HRSI tile, weighing only 12lb/cu ft. It derives from adding an alumino-borosilicate fibre called Nextel to the pure silica of the standard HRSI tile. In addition to being lighter it provides greater tensile strength and can protect against temperatures 100° greater than with the standard tile.

The remarkable properties of all these HRSI tiles allow them to be held by the edges in an ungloved hand within seconds of being soaked at 2,300°F. They are coated on the top and sides with a mixture of powdered tetrasilicide and borosilicate glass, sprayed on to a coating of about 0.16–0.18mm. This provides a waterproof coating with a shiny black surface, which sheds about 95 per cent of the heat encountered, the substance of the tile dissipating the heat, reducing the temperature from nearly 2,300°F to well below 250°F within a maximum depth of 5 inches.

LRSI tiles are essentially the same but thinner and approximately 8in x 8in square, with a white 0.1mm-thick optical coating made from silica compounds with shiny aluminium oxide properties applied to the top and sides. They are attached to the upper surfaces of both wings and to substantial areas on the slab-sided fuselage and vertical tail. The white coating is intentional to help thermal management of the Orbiter's temperature in space.

Developed after the assembly and roll-out of OV-102 Columbia, Advanced Flexible Reusable Surface Insulation (AFRSI) has been used to replace most of the white LRSI tiles. It consists of a low-density fibrous silica batting made up from a combination of high-purity silica and 99.8 per cent amorphous silica, sewn with a silica thread to give it a quilted appearance. To protect areas where temperatures are never likely to exceed 700°F, a Nomex Felt Reusable Surface Insulation (FRSI) is bonded directly to the Orbiter.

The total weight of the thermal protection system varies slightly with each Orbiter but the average for the total vehicle is 8,574lb, of which 20 per cent is the small quantity of RCC covering just 2 per cent of the surface. Just over 50 per cent of the total weight is accounted for in black HRSI and a further 12 per cent on white-coated LRSI, and combined these cover 66 per cent of the surface area of the Orbiter. The balance is made up from FRSI and various small quantities of other temperature control materials.

There are more than 27,000 tiles on each Orbiter, each assigned a unique position in relation to the rest. Each tile has a number

LEFT The compound curvature of the forward wing leading edge and fuselage fillet create challenges for tile mappers and fixers. Note the cargo bay door with felt insulation. (NASA)

LEFT Lifting into place RCC panels for the wing leading edge (see page 66) to which they are firmly bolted and will provide the aerodynamic curvature. (NASA)

BELOW The RCC panels are crucial to the survival of the Orbiter during re-entry. It was damage to one of these that doomed Columbia, opening a path for hot gases to penetrate the wing and melt the structure. (NASA)

stamped on it in yellow and should a replacement be necessary a computer can manufacture another and apply the identical location number so that technicians can place it precisely on the Orbiter. The flags and letters are a silicon-based material coloured with pigments and sprayed onto the exterior, essentially the same paint as that used for car engines and capable of withstanding 1,000°F.

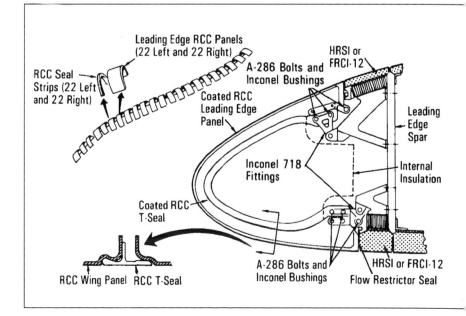

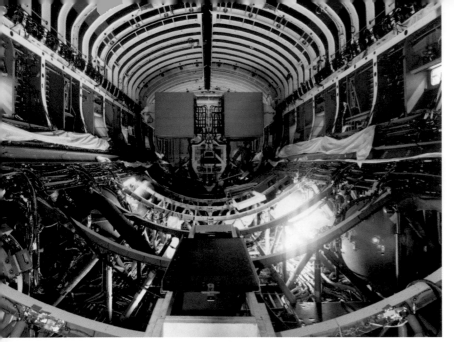

Electrical power system

The means to keep the Shuttle operating from just before launch until shortly after landing comes from an electrical system provided by fuel cells. This is a proven technology that dates back to the mid-1960s and has been used to provide electrical energy for Gemini and Apollo spacecraft before the Shuttle. Belying the reference to 'fuel', the system works by a form of reverse electrolysis. Instead of using an electrical current to break

down water into its constituent molecules of hydrogen and oxygen, those two chemicals are brought together over a catalyst to produce electrical energy and water as a bi-product. In this way the one system manufactures water vital for the spacecraft cooling system, for reconstituting food and, when purified, for drinking purposes.

The hydrogen and oxygen 'fuels' are known as reactants and while both are needed to power the fuel cells, the same oxygen tanks can also be used to provide part of the pressurised crew compartment with its life supporting atmosphere, mixed with nitrogen from another tank. This economy of scale, integrating several vital functions into one contained system, is efficient and saves a great deal of weight over parallel systems that are each dedicated to a specific function.

Gemini 5 was the first spacecraft to run on this technology in August 1965, operating for eight days on fuel cells. Over the intervening decades fuel cell technology has improved considerably and the small, largely experimental units of the 1960s are today a highly reliable and efficient form of power. Other electrical sources such as batteries or solar cell arrays are too big or cumbersome for the amount

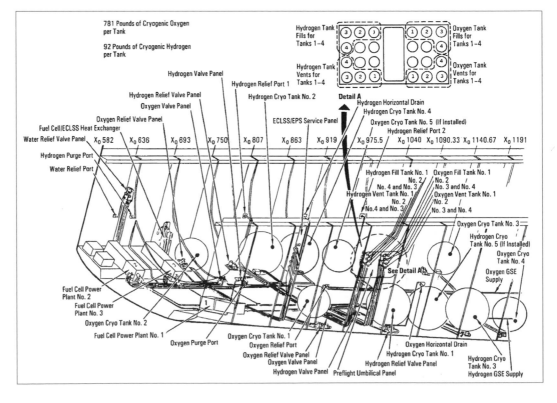

of power required to operate the Shuttle. The only alternative would be nuclear isotopes of plutonium 238, producing power from thermocouples that use heat from radioactive decay to produce electricity. But the amount of shielding, and the levels of power demanded by the Shuttle, renders that option unsuitable.

To power all of its systems the Shuttle needs a continuous supply of 21kW at 28 volts dc, with a surge capacity to draw down a maximum 36kW for 15-minute peaks. To provide this the Shuttle has three fuel cells, each producing 7kW continuous power and 12kW at peak. Before launch all electrical power is provided by a ground supply through an umbilical into the Orbiter up to 3 minutes 30 seconds before lift-off, when electrical power supplies switch to the fuel cells. The electrical distribution system carries the 28 volt dc supply to three main dc busses. Each of these supplies power to three solid-state, single-phase inverters which constitute one three-phase alternating current bus. The nine inverters convert the dc power to 115 volt, 400-hertz, ac power for distribution to three ac buses for the Orbiter's ac loads. This system not only provides power to the Orbiter systems but also to experiments and equipment in the payload bay and, while they are attached, to the two Solid Rocket Boosters and the External Tank.

The reactants are stored in spherical pressure vessels as cryogenic fluids, because stored as gases they would require more volume than the entire size of the Orbiter itself! So they must be stored as liquids and take up only a tiny fraction of that volume, but oxygen boils at –297°F and hydrogen at –425°F so they are held in spherical tanks under great pressure. Essentially thermos flasks for keeping heat out, the tanks are paired in sets installed in the mid-fuselage between the U-shaped frames and beneath the payload bay liner. Up to five sets are installed for long duration flights.

The 201lb oxygen tanks are 33.4in in diameter with a volume of 11.2cu ft and hold 781lb of heavy oxygen. The 216lb hydrogen tanks are 45½in in diameter with an internal volume of 21.4cu ft, which can hold 92lb of the light hydrogen. Each pressure vessel is actually two-in-one; the vacuum between the two shells ensures that no heat leaks in, while the innermost shell is bolstered by 12 low-conductive supports to

minimise thermal transfer. The reactants inside each tank are gradually depleted as they are consumed by the fuel cells or, in the case of the oxygen, by the environmental control system as well. Constant pressure must be maintained for the liquids to be expelled through the pipes connected to the outlet, but as the volume of liquid goes down within the finite volume of each tank, the pressure drops. Heaters inside each tank ensure that this does not happen. As the liquids are consumed, the temperature is raised by a few degrees to begin the process of converting the liquids into gases, hence maintaining the pressure.

These heaters can operate in either automatic or manual mode, with a relief valve opening if pressure in the oxygen tank rises above 1,005lb/sq in, or in the hydrogen tank to 310lb/sq in. The reactants flow through relief valve and filter packages to the fuel cells – oxygen at 815–881lb/sq in and hydrogen at 200–243lb/sq in. The reactant tanks are loaded on the launch pad and are topped up until the point when the ground equipment is automatically disconnected at 2 minutes 35 seconds before lift-off.

To operate the fuel cell, oxygen flows to the oxygen electrode and reacts with the water, returning electrons to produce hydroxyl ions, which migrate to the hydrogen electrode and enter the hydrogen reaction. The hydrogen is routed to the hydrogen electrode, reacting with the hydroxyl ions from the electrolyte. This electrochemical reaction over a potassium hydroxide electrolyte produces the electrons for

ABOVE NASA has a history of pioneering the use of compact hydrogen-powered fuel cells for electrical energy production. All its manned spacecraft since Gemini (1965–66) have used this technology. With a length of 40in, each of the Shuttle's three 200lb fuel cells has a width of 15in and a height of 14in. *(United Technologies)*

electrical power, water and heat; the amount of oxygen and hydrogen consumed is proportional to the amount of power demanded.

Excess water vapour is removed by an internal circulating system and mixed with replenishing hydrogen before it is diverted to a condenser, where waste heat from the system is transferred to the fuel cell coolant system. A water separator extracts the liquid water, which is sent to a potable water storage tank on the lower deck of the crew compartment. From there it can be sent to the Orbiter cooling system, or it can be used by the crew for reconstituting food or for drinking.

During normal operation the reactants are totally consumed in the process of producing electricity, but some inert gases and contaminants can collect to reduce the efficiency of the fuel cells. To prevent this becoming excessive, twice a day the crew purges the fuel cells. They do this by opening purge valves to allow the oxygen and hydrogen to flow in an open loop manner, circulating reactants through the stack and blowing contaminants out through purge lines and vents to space.

BELOW The fuel cell units are situated in the extreme forward part of the fuselage centre section, with the reactant vessels to the rear. The six fill and drain ports connect to ground service equipment at launch and landing sites. *(NASA)*

The three fuel cells are each 14in high by 15in wide and 40in long, and each weighs approximately 255lb. In normal operation each is rated at from 2kW at 32.5 volts dc and 61.5 amps to 12 kW at 27.5 volts dc and 436 amps. Average power consumption is 21kW, of which 14kW is taken by the Orbiter for its systems and 7kW for the experiments on board, or for payloads in the cargo bay. The fuel cells are serviced between flights and returned to use until they have each accumulated about 2,000hr of service time on line.

Environmental control and life support

The Orbiter is designed for supporting up to eight crewmembers for up to 21 days, but usually flies for a 14-day period before returning to earth. Great demands are placed upon the Shuttle compared to all previous spacecraft designed to carry humans beyond earth's atmosphere. From the first US-manned space flight in May 1961 to the joint docking flight with the Russians in July 1975, Mercury, Gemini

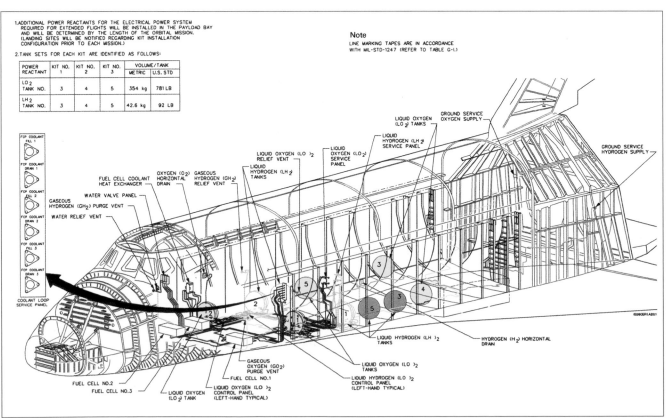

and Apollo spacecraft had a pure oxygen atmosphere pressurised to little more than one-third sea-level pressure. The Shuttle creates a breathable environment duplicating that on the earth's surface. To do that it must replicate the earth's atmosphere using integrated systems that are not only highly efficient but reliable – human lives depend upon it.

Two-gas cabin atmosphere

The choice of a two-gas atmosphere at sea-level pressure is made not only for comfort but for physiological reasons. The original purpose of the Shuttle was to help assemble, and keep supplied, a fully-manned space station capable of hosting astronauts for up to one year in orbit. The Shuttle itself was designed to be the world's first fully reusable space ferry and both planners and engineers considered it to be more akin to a cargo plane than a spacecraft. From the outset, it was important to make it compatible with the space station for which it was built.

Humans need oxygen for the healthy working of the brain, but they can do without nitrogen for no more than three or four weeks. Because of that, pure oxygen flights with early manned spacecraft were limited to two weeks in duration. The longer flights made possible by the Shuttle made an oxygen/nitrogen atmosphere essential. To achieve this two-gas atmosphere the Shuttle benefits from combining the oxygen supply with that of the fuel cell reactant system, with an additional supply of nitrogen to add to the atmosphere in the pressurised crew compartment for the 20/80, oxygen/nitrogen mix. This helps keep weight down and reduces potential problems by reducing the number of components.

The Shuttle's environmental control and life support system, or ECLSS, has responsibility for maintaining and revitalising the air system in the pressurised crew compartment, supporting operations with the airlock, maintaining proper control of temperatures throughout the Orbiter, supplying water for the crew and the coolant system and controlling a waste management system. It is not only responsible for keeping the crew alive and comfortable, but for conditioning all the unpressurised areas of the orbiter, too.

Measurement of the atmospheric gases in

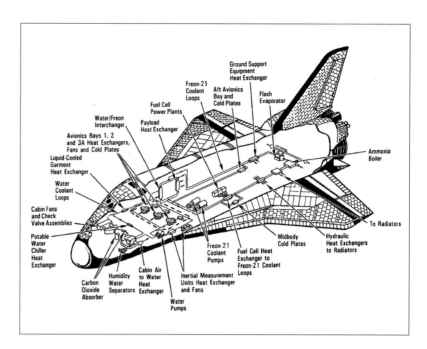

ABOVE With so much electrical equipment on board, thermal control and heat removal is vital. The relevant coolant rails and cold plates are distributed throughout the structure. *(NASA)*

BELOW The equipment bay is located below the floor of the mid-deck, and contains key components of the environmental control system. In this view, the aft bulkhead is at the bottom and the nose is at top. *(NASA)*

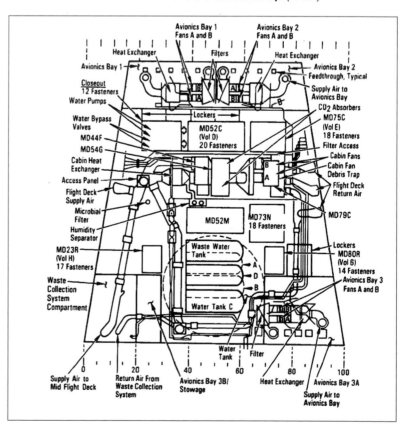

the crew compartment ensures that the oxygen level is maintained between 2.95lb/sq in and 3.45lb/sq in and that the nitrogen partial pressure remains at 11.5lb/sq in. The 14.7lb/sq in atmosphere is regulated by monitoring the oxygen level, since that is critical for proper functioning of the human brain and must never be allowed to fall below the lower limit of that band. Generalised – or tissue – hypoxia can result, which at best damages cells and can cause brain damage or death at worst. The oxygen supply comes from the same tanks used for fuel cell reactants in the electrical system. It is delivered to the ECLSS as a gas at room temperature and a pressure of 835–853lb/sq in.

The gaseous nitrogen supply consists of two systems each with two tanks. Each tank is pressurised to 2,964lb/sq in at a temperature of 80°F. An auxiliary gaseous supply tank containing 67.6lb of oxygen can be installed for some missions. It is serviced at 2,440lb/sq in to provide high flow rates along with the gaseous nitrogen and would maintain a cabin pressure of 8lb/sq in. As said, the oxygen comes from the shared system with the fuel cell reactant, while the nitrogen supply not only provides gas for the crew compartment atmosphere but also to pressurise the potable and waste water tanks and for re-pressurising the airlock. Losses due to cabin gas escaping into space and human breathing amount to about 7.7lb of nitrogen and 9lb of oxygen per day. The two primary and two secondary nitrogen supply tanks are fabricated from filament-wound Kevlar fibre with a titanium liner, and each tank has a volume of 8,181cu in. Both tanks in each system are connected by a manifold.

Based on the crew compartment volume of 2,300cu ft, and a flow rate of 330cu ft/min of mixed nitrogen/oxygen air, the entire pressurised area gets a complete change of air every seven minutes. Five independent air loops are provided, for the crew compartment itself, for three avionics bays and for the inertial measurement units, which are the core of the guidance and navigation section. Air that passes through the cabin picks up moisture, odour, carbon dioxide and micro-debris, which is drawn through a 300-micron filter by one of two cabin fans. The air is then ducted to two

lithium hydroxide canisters where the harmful CO_2 is removed.

This 'scrubbing' of the air is important for maintaining alertness and mental health. Even partially toxic environments cause deterioration of brain efficiency and lack of concentration, sometimes causing nausea and sleepiness. The lithium hydroxide canisters are situated under the floor of the mid-deck and can be accessed by the crew and changed alternately every 12 hours. For a nominal crew of seven astronauts, these canisters are changed every 11 hours with replacement canisters stored under the mid-deck floor between the cabin heat exchanger and the water tanks.

Cabin temperature regulation

A water coolant loop system provides a regulated cabin temperature and transfers heat to the Freon-21 coolant loop interchanger, helping keep the three avionics bays cool where some electronics are mounted on cold plates. These transfer heat conducted from avionics equipment to the radiators, or to the flash evaporator if the payload bay doors have not been opened. The flash evaporator operates during launch and ascent to orbit and during re-entry and controls temperature through boiling the water. The radiators located on the inner face of the payload bay doors are capable of rejecting up to 21,500BTU/hr or for missions where a lot of heat is likely to be generated by some equipment, an additional radiator panel can be attached to the aft underside of the aft right and left doors and this raises rejection capacity to 29,000BTU/hr.

The radiator panels are constructed on aluminium honeycomb sheets 10½ft wide and 26.7ft long. The forward radiator panels are two-sided and have a core thickness of about 0.9in. Each of the forward deployable radiators have 68 tubes spaced 1.9in apart through which the Freon-21 flows at rates proportional to the demand of the heat rejection system. The aft radiator panels are fixed down to the inner face of the payload bay doors and contain 26 longitudinal tubes at 4.96in intervals. Each radiator panel is 10ft wide and 15ft long and together provide an effective surface area of 1,195sq ft. The tubing which carries the Freon coolant is more than one mile long.

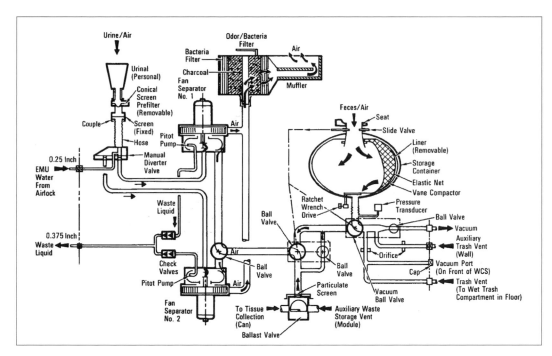

Water system and toilet facilities

Supply and waste water systems serve the flash evaporator, the crew's water and hygiene requirements, and the supply system stores water produced by the fuel cells. Four supply tanks and one waste water tank are situated under the mid-deck floor in the crew compartment. Four potable water tanks have a usable capacity of 165lb and each is 35½in in length and 15½in in diameter with an empty weight of 39½lb. The three fuel cells produce up to 25lb of potable water every hour which flows to a water relief valve and from there to the potable tanks. The waste water tank is the same size and has the same capacity as each of the four potable tanks and is pressurised by gaseous nitrogen.

Going to the toilet in space can prove difficult, but it is much easier now that astronauts have reasonably well engineered facilities in the Shuttle under the designation of 'waste collection system' or WCS. Located in the mid-deck area the WCS is a cubicle 2ft 5in wide just aft of the side hatch. The toilet consists of a commode, for collection, storage and processing of body wastes. Fecal waste is dried and stored and can be of medical use in evaluating the health of the crew, but the urine is transferred to the waste water tank. When not in use the commode is depressurised for drying fecal matter and biologically deactivating it.

Use of the commode in weightlessness is made that much easier by foot restraints, in the form of toe bars, and a handhold each side of the lid. A flush handle is located to the user's right. The seat is shaped to accommodate the buttocks and the commode interior is lined with a porous bag liner, while a fan separator extracts air from the liquid and passes it through an odour scrubber. Filters strip the air molecules of odour and bacteria before returning them to the cabin recirculation system as part of the ECLSS.

The urinal is designed to accommodate the physical differences between male and female astronauts, and essentially consists of a pipe with alternative fixtures at the end for passing urine. It can be used standing up. If the fecal waste system breaks down, fecal bags almost identical to those used in Apollo spacecraft are used, with adhesive tape for application to the buttocks, a seal and storage system being provided.

Airlock module

The last subsystem addressed by the environmental control and life support system is the airlock module, which can be attached before launch to the rear of the mid-deck area or outside in the forward part of the payload bay. Either way, it provides the ability for two suited astronauts to move outside without depressurising the crew compartment. This is vital because the astronauts use pure oxygen

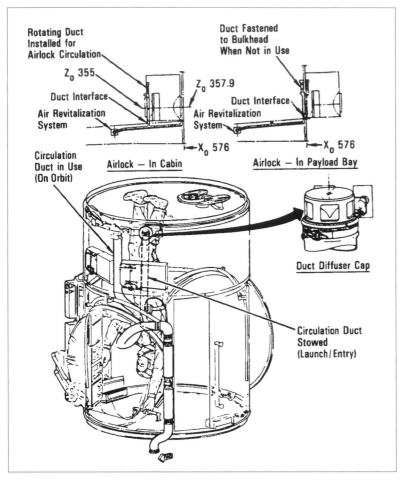

Rotating Duct
Installed for
Airlock Circulation

Z_0 355

Duct Interface

Air Revitalization
System

Circulation
Duct in Use
(On Orbit)

Z_0 357.9

Duct Fastened
to Bulkhead
When Not in Use

Duct Interface

Air Revitalization
System

X_0 576

X_0 576

Airlock — In Cabin

Airlock — In Payload Bay

Duct Diffuser Cap

Circulation Duct
Stowed
(Launch / Entry)

in their suits and breathe the mixed gas atmosphere of oxygen and nitrogen inside the Orbiter. Ideally the astronaut would breathe the same air in the suit as in the Orbiter, but with a requirement for at least 2lb/sq in of oxygen, the balance of nitrogen would demand an internal suit pressure of 10lb/sq in, which is too much for a suit to remain flexible and not become stiffened and rigid. Even so, the space suit is

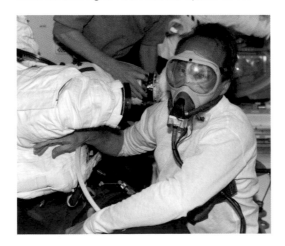

pressurised to 4.7lb/sq in, which provides a safe 3lb/sq in for breathing plus 1.7lb/sq in to compensate for carbon dioxide and water vapour pressure, leaving a sufficient alveolar pressure for the body. This is roughly equivalent to the oxygen a walker breathes in when climbing a 6,000ft mountain. Because of that, the work load of the astronaut performed during an EVA must be adjusted so as to avoid fatigue, which even a walker at 6,000ft would feel!

The airlock module has an interior diameter of 5ft 3in and is 6ft 11in in length, with two 40in diameter D-shaped hatchways 3ft across, the innermost having a 4in diameter window for observing the occupants from the mid-deck area. There are two pressure-sealed hatches on opposite sides, one for getting into the airlock and the second for getting out into the area of the payload bay after the module has been depressurised. Each hatch has a double pressure seal for safety and a leak-check disconnect is located between the hatch and the pressure seal to ensure that it is truly airtight. If the airlock module is attached to the outside, in the payload bay and against the crew compartment bulkhead, the exterior carries thermal insulation to prevent the cold of space being conducted to the interior. Pressurisation and re-pressurisation of the airlock module is controlled from the mid-deck area.

For a space-walking astronaut preparing for an EVA there are several hand holds and four floodlights to illuminate the interior. It is inside the airlock that an astronaut dons the space suit, or Extra-vehicular Mobility Unit (EMU), which has an upper torso section held in a frame for ease of donning. Each EMU weighs 225lb. To prepare for an EVA the pressure in the crew compartment is allowed to bleed down to 12.5lb/sq in the day before an EVA is scheduled to begin. For 45 minutes before the space walk the astronaut breathes 100 per cent pure oxygen through a face mask connected to a delivery pipe, and the pressure of the crew compartment is further reduced to 10lb/sq in. This is so as to remove nitrogen from the blood and prevent the bends.

To prepare for going outside the crewmember dons a Liquid-Cooled Garment (LCG), which will be connected to the backpack so that water can flow through tubing sewn into

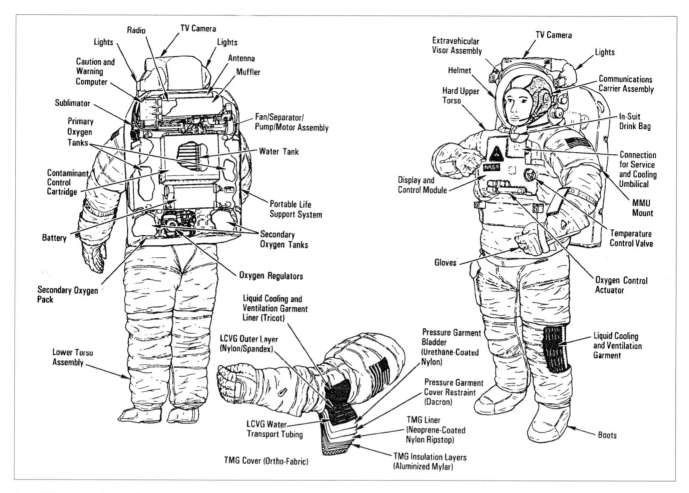

Labels (left figure): Radio, Lights, TV Camera, Lights, Caution and Warning Computer, Antenna, Muffler, Sublimator, Fan/Separator/Pump/Motor Assembly, Primary Oxygen Tanks, Water Tank, Contaminant Control Cartridge, Portable Life Support System, Battery, Secondary Oxygen Tanks, Oxygen Regulators, Secondary Oxygen Pack, Liquid Cooling and Ventilation Garment Liner (Tricot), Lower Torso Assembly, LCVG Outer Layer (Nylon/Spandex), LCVG Water Transport Tubing, TMG Cover (Ortho-Fabric), Pressure Garment Bladder (Urethane-Coated Nylon), Pressure Garment Cover Restraint (Dacron), TMG Liner (Neoprene-Coated Nylon Ripstop), TMG Insulation Layers (Aluminized Mylar)

Labels (right figure): Extravehicular Visor Assembly, TV Camera, Helmet, Lights, Communications Carrier Assembly, Hard Upper Torso, In-Suit Drink Bag, Connection for Service and Cooling Umbilical, Display and Control Module, MMU Mount, Temperature Control Valve, Gloves, Oxygen Control Actuator, Liquid Cooling and Ventilation Garment, Boots

the LCG and maintain the temperature of the body at acceptable levels. Then the astronaut moves into the open airlock where the EMU is donned and oxygen pressure and suit flow tested for leaks.

Communications is first maintained through connectors in the airlock module and then through the backpack which, along with the suit, is put on inside the airlock module. It is necessary for the astronaut to pre-breathe oxygen on the suit loop for a further 40 minutes before the process of leaving the spacecraft can begin.

Auxiliary Power Unit (APU)

Because the Orbiter performs the function of an aircraft during its descent through the atmosphere, it has flying control surfaces and landing gear and many of these functions require hydraulic systems to operate them. The Orbiter also has three very powerful main engines,

which are controlled by the hydraulics as well as the gimbaling mechanisms which drive the Shuttle on its course during ascent. Power for the hydraulics comes from three Auxiliary Power Units (APUs), much like those on commercial aircraft that drive the onboard systems when the aircraft's engines are not running.

In the Shuttle, the APUs are situated in the aft-fuselage of the Orbiter and operate on liquid

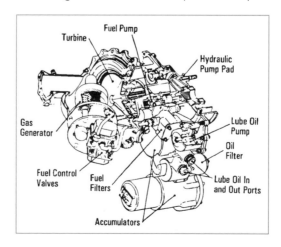

Labels: Turbine, Fuel Pump, Hydraulic Pump Pad, Gas Generator, Lube Oil Pump, Oil Filter, Fuel Control Valves, Fuel Filters, Lube Oil In and Out Ports, Accumulators

ABOVE The extravehicular suit assembly is an evolution of the suit worn on the moon by Apollo astronauts, with an autonomous backpack for Extra Vehicular Activities (EVAs) lasting up to eight hours. (NASA)

LEFT Hydraulic power for the Shuttle is provided by three Auxiliary Power Units (APUs). Before lift-off and afte re-entry, the APUs provide power for the gimbals on the three main engines, the aerodynamic control surfaces and the landing gear. (NASA)

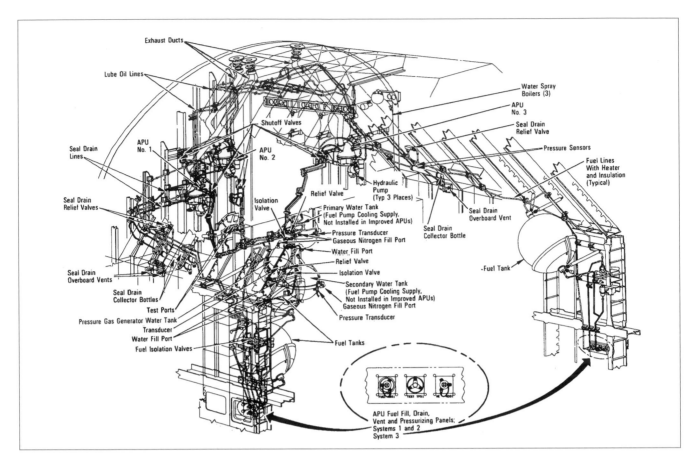

Labels in image:

Exhaust Ducts
Lube Oil Lines
Shutoff Valves
Water Spray Boilers (3)
APU No. 3
Seal Drain Relief Valve
Pressure Sensors
Seal Drain Lines
APU No. 1
APU No. 2
Relief Valve
Hydraulic Pump (Typ 3 Places)
Fuel Lines With Heater and Insulation (Typical)
Seal Drain Relief Valves
Isolation Valve
Primary Water Tank (Fuel Pump Cooling Supply, Not Installed in Improved APUs)
Pressure Transducer
Gaseous Nitrogen Fill Port
Water Fill Port
Relief Valve
Isolation Valve
Secondary Water Tank (Fuel Pump Cooling Supply, Not Installed in Improved APUs)
Gaseous Nitrogen Fill Port
Pressure Transducer
Seal Drain Overboard Vent
Seal Drain Collector Bottle
Fuel Tank
Seal Drain Overboard Vents
Seal Drain Collector Bottles
Test Ports
Pressure Gas Generator Water Tank
Transducer
Water Fill Port
Fuel Isolation Valves
Fuel Tanks
APU Fuel Fill, Drain, Vent and Pressurizing Panels; Systems 1 and 2 System 3

hydrazine fuel delivered to separate pumps, one on each APU. A gas generator decomposes the fuel through a catalyst producing hot gas, which drives a two-stage turbine. In turn, the gases are vented directly overboard though an exhaust duct in the upper section of the aft-fuselage assembly. The turbine provides the mechanical power through a shaft, which drives gears in a gearbox. This drives a fuel pump, hydraulic pump and lubrication oil pump. The oil pump provides pressure to the hydraulic system and drives the various control surfaces, much like those on a conventional aircraft.

The three APUs are fired up five minutes before lift-off to position the main engines for ignition, for the various propellant valves, and to position the elevons and rudder on the Orbiter. The APUs are shut down after the Shuttle reaches orbit and one of these will be fired up a day before re-entry to demonstrate that the Orbiter has at least one system fully operational. The two remaining APUs are fired up after the de-orbit burn and will be used to control the flight control surfaces, deployment of the landing gear and doors and for properly positioning the three main engines in the gimbals after rollout to a stop.

The three fuel tanks for the three APUs are supported on cantilevered struts in the aft-fuselage and hydrazine is held inside a flexible diaphragm inside each tank. To squeeze fuel out of the rigid spherical 28in diameter tank, nitrogen gas is introduced under pressure to the space between the wall of the tank and the diaphragm, in what is known as positive-expulsion. Fuel is driven by the APU pump at 1,400–1,500lb/sq in and each gas generator consists of a bed of Shell 405 acting as a catalyst inside the exhaust chamber. It decomposes the hydrazine into a hot gas at 1,700°F and exhausts the products at a temperature of 1,000°F. The turbine spins at 74,160rpm for normal conditions but can be 'throttled' to a maximum 82,800rpm.

Use of the APUs fitted in the first decade of Shuttle operations revealed an unexpected need for major refurbishment after 20 hours but a more advanced design was incorporated by 1990, promising 75 hours of use before replacement.

Hydraulic system

The three APUs exist to provide hydraulic power via three pumps, one to each APU, providing the pressure for respective hydraulic systems. Hydraulic pressure is usually maintained at 2,900–3,100lb/sq in. Each hydraulic pump usually operates at a displacement rate up to 63gal/min at 3,000lb/sq in. The hydraulic fluid is a synthetic hydrocarbon to reduce the risk of fire and is to specification MIL-H-83282.

The aero-surfaces on the Orbiter, the elevons, rudder/speed brake and the body flap, are all hydraulically operated. Each elevon on the trailing edge of the wing surfaces can be operated by any one of the three systems. The rudder/speed brake is driven by six hydraulic motors in a special power drive unit, three for the rudder and three for the speed brake function. Each motor is supplied by a different hydraulic system, and so the loss of one will result in the loss of one motor only.

Hydraulic actuators are also used for several functions on the main engines, controlling oxidiser and fuel pre-burner valves, main oxidiser and main fuel valves, and for operating the SSME gimbals, which align the thrust correctly for trajectory control. When the External Tank is separated, the two umbilicals that had connected the Orbiter to the ET are retracted by the hydraulics and after descending down through the atmosphere at the end of the mission the landing gear doors are opened, and the wheel assemblies lowered, by hydraulic power.

Landing gear

The Orbiter has been designed with a standard tricycle gear configuration, with each landing gear incorporating a shock strut and two wheel and tyre assemblies. The nose gear is retracted forward and up into the lower section of the forward fuselage and enclosed by two doors. The main gear is retracted forward and up into the left and right lower wing areas. These retractions take place in the Vehicle Assembly Building after the Orbiter has been towed round from the Orbiter Processing Facility for mating with the External Tank and the Solid

LEFT The landing gear nose leg on an Orbiter in the Vehicle Assembly Building, where it will be manually raised and locked and the doors closed before the Orbiter is rotated for mating with the External Tank. *(NASA)*

BELOW The design philosophy of the landing gear is simplicity, for reliability and safety, but if the hydraulics fail to work it can be lowered by gravity. *(NASA)*

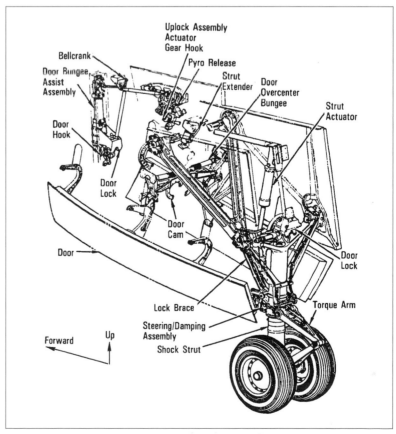

ABOVE Considerable improvements have been made to the wheel assemblies for the Orbiter, earlier assemblies leaving a lot to be desired, which resulted in tyre damage on more than one flight. *(NASA)*

RIGHT The landing gear doors are thermally insulated and are lightweight structures. They must seat correctly, providing a thermal seal between the interior of the Shuttle and the hot gases of re-entry. *(NASA)*

Rocket Boosters. They are done manually and there is no provision for gear retraction because the Orbiter is a glider and cannot elect to go around for another landing attempt.

Landing gear is deployed after the Orbiter has an indicated airspeed of no greater than 300kts (345mph) and an altitude of about 250ft. The main shock strut is the primary source of attenuating the landing impact, these having air/oil shock absorbers to damp the rate of compression. Each main gear wheel has an electro-hydraulic disc brake with anti-skid control and brakes are operated by the commander from the left seat. The nose gear

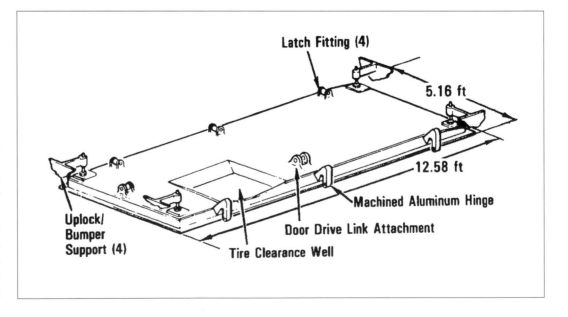

carries a hydraulic steering actuator, which can be controlled through the general-purpose computers or the pilots' rudder pedals.

The nose landing gear tyres are 32in by 8.8in with an inflation pressure of 300lb/sq in and a burst pressure three times that. The tyres are inflated with gaseous nitrogen and the nose gear has a stroke of 22in. The main gear tyres are 44½in by 16in and 21in with an inflation pressure of 315lb/sq in and the shock strut has a stroke of 16in. The maximum allowable sink rate at wheel contact is 6ft/sec for an Orbiter weighing 212,000lb, or 5ft/sec for a 240,000lb Orbiter at landing.

Each main landing gear wheel has a disc brake consisting of nine discs, four rotors, three stators, a back plate and pressure plate. The carbon-beryllium rotors are splined to the inside of the wheel and rotate with the wheel itself. The carbon lined stators are splined to the inside of the axle assembly and do not rotate. Many improvements have been made to the wheels, brakes and materials fitted to the Orbiter with high-energy absorbing designs replacing the early fixtures.

ABOVE A view inside the main landing gear bay with thermal insulating tiles being applied to the undersurface of the wing. *(NASA)*

LEFT When is a door more than a door? Answer – on the most vulnerable part of the Shuttle, where a thermal leak of hot gases between the nose cap and the landing gear doors would be fatal in this, the hottest part of the Orbiter during re-entry. *(NASA)*

Chapter Four

Thrust for the Shuttle

The design of the Orbiter is dictated by the need for it to support seven crewmembers for two weeks in space, and to safely return its crew and cargo to earth. In providing for those requirements, it benefits from aircraft design, manufacturing and assembly expertise, but where it becomes a spacecraft is in the rocket motors that dictate where it goes and what it does in orbit. The Shuttle has power on a scale never before put into an airframe, and propulsion systems that test the very limits of technology.

LEFT Propulsion for getting into space is down to the three Space Shuttle Main Engines attached to the thrust structure in the aft-fuselage assembly. The two Orbital Manoeuvring System motors provide thrust for nudging the Shuttle into orbit, for rendezvous manoeuvres in space and for de-orbiting the Shuttle at the end of a mission. Note the Ku-band antenna on the starboard side and the remote manipulator arm on the port side. *(NASA)*

RIGHT The Shuttle is embraced by service masts and a crew access arm terminating in the box-like 'white room' where the crew gains access to the Orbiter. *(NASA)*

The very essence of good aeronautical design is to follow well established guidelines, principles and theories of flight and flying. The Shuttle had no precursor and therefore had no precedent to follow; and being unique it had no parallel with which its design could be compared. Instead, the configuration had been chosen to satisfy financial and military requirements, both of which severely compromised its design. And in turn that reduced its capabilities and its performance. Contravened, too, was the dictum of sound aerospace practice, never to develop a completely new propulsion system for a new and untried airframe. When the configuration of the Shuttle was fixed in early 1972 it brought a unique design style for aerospace engineers to work with, and that had its own range of very special problems.

Known as a parallel-burn, thrust assisted Orbiter concept, the Shuttle would carry its own engines in the rear thrust structure of the aft fuselage, burning propellants carried in the large cylindrical structure to which it was attached, incorporating separate tanks for liquid oxygen and liquid hydrogen. For the first two minutes the Shuttle would be assisted on its way by two massive solid rocket boosters, dwarfing anything of this type flown before. Never had such a combination of rocket motors and boosters been built like this and nothing like the Orbiter had ever flown into space, let alone returned to a conventional runway for use many times over.

The engineering challenges were immense and would cut new ground, but in one respect the Shuttle got a head start. The outstanding

RIGHT The original concept for a two-stage fully reusable Shuttle, involved 12 rocket motors in the fly-back booster and two similar motors in the Orbiter. Conventional (air-breathing) jet engines were to have been provided for go-around capability and for ferry flights back to the launch site. *(NASA)*

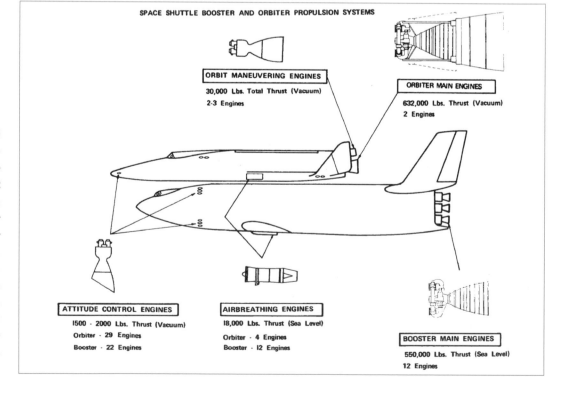

SPACE SHUTTLE BOOSTER AND ORBITER PROPULSION SYSTEMS

ORBIT MANEUVERING ENGINES
30,000 Lbs. Total Thrust (Vacuum)
2-3 Engines

ORBITER MAIN ENGINES
632,000 Lbs. Thrust (Vacuum)
2 Engines

ATTITUDE CONTROL ENGINES
1500 - 2000 Lbs. Thrust (Vacuum)
Orbiter - 29 Engines
Booster - 22 Engines

AIRBREATHING ENGINES
18,000 Lbs. Thrust (Sea Level)
Orbiter - 4 Engines
Booster - 12 Engines

BOOSTER MAIN ENGINES
550,000 Lbs. Thrust (Sea Level)
12 Engines

management techniques that brought success to NASA's moon-landing Apollo programme, and the tools developed by industry to co-ordinate several thousand subcontractors and component manufacturers, smoothed the way for what was arguably the most difficult aerospace project of all time. When engineers moved into detailed design of the Shuttle it had been only 25 years since 'Chuck' Yeager had smashed the sound barrier. Now, at the pinnacle of aircraft design, a flying machine would reach 17,500mph and remain in space for up to three weeks. As personnel moved from the Apollo programme to the Shuttle a rich harvest of talent, engineering capability and experience provided the Shuttle with a running start, where its predecessor had struggled to build methodologies in work and in ways of running big space projects.

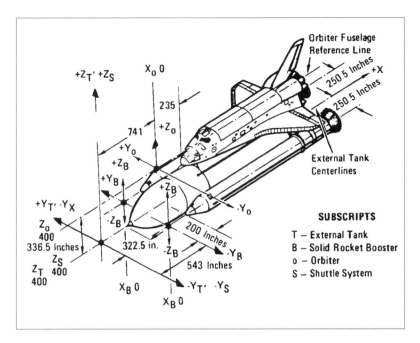

Systems integration

When the Second World War ended in 1945, US aircraft and engine industries met new challenges in performance and capability through a completely different approach to design and engineering. Before then, aircraft were built to a general design,

each element being taken to a different design office and given to engineers and draughtsmen who would take the specific shape of the aircraft and complete the detailed design on wing, fuselage, tail or landing gear through separate teams of engineers. Only then would it all be brought together. But the new Weapons System concept adopted by the US military in the early 1950s integrated all the various

ABOVE The coordinate system for the Shuttle is based around the conventions used in the aircraft industry for referencing pitch, roll and yaw. The very precise positioning of Orbiter, External Tank and SRBs has a great effect on aerodynamic performance during ascent. *(NASA)*

LEFT In the Vehicle Assembly Building (VAB) at the Kennedy Space Center, a specially designed sling is manoeuvred into position for rotating the Orbiter 90° prior to mating it to the External Tank. *(NASA)*

elements of an aircraft, missile or engine into one homogeneous whole. It was never that way with any of the pre-Shuttle US manned space vehicles, where Mercury, Gemini and Apollo all utilised launch vehicles developed separately.

More than any other aerospace vehicle, the Shuttle is a marvel of systems integration made possible only through an interdisciplinary approach balancing all four elements: Space Shuttle Main Engine (SSME); External Tank (ET); Solid Rocket Booster (SRB); and the aero-thermal design characteristics of the Orbiter. Each is dependent on overriding design, manufacturing and flight profiles to which all the primary Shuttle elements conform. For example, two fundamental parameters dictated the design of the Orbiter itself and of the assembled configuration of propellant tanks and boosters: a need to limit maximum atmospheric pressure during ascent to 650lb/sq ft; and an upper limit of 3g on the force of acceleration going up into orbit and returning through the atmosphere.

Known as 'maximum dynamic pressure', the peak of atmospheric pressure is the point at which the rapidly accelerating Shuttle receives the greatest pressure from the

earth's atmosphere. Sitting on the launch pad, the ambient atmosphere exerts a pressure of 14.7lb/sq in (or 2,117lb/sq ft), but as it accelerates through the atmosphere that pressure on all the leading edges of the vehicle goes up. The dynamic pressure measured on the vehicle is the difference between the static ambient pressure and that imposed as the vehicle pushes forward. In reality the atmospheric pressure goes down as altitude is gained and the air gets thinner. But even at a maximum 3g, the acceleration of the Shuttle is such that it achieves that maximum dynamic pressure of 650lb/sq ft, a value known to aerodynamicists as 'Max-q', before the earth's atmosphere has begun to tail off with altitude.

The max-q of 650lb/sq ft was imposed for structural design reasons and all elements of the Shuttle were built with that force as the ceiling to which they would be imposed in flight. To achieve that, the thrust of the Orbiter's three main engines and the two solid rocket boosters had to be controlled so that the Shuttle, getting lighter as propellant was consumed, would not run away with itself and impose destructive forces on the forward end. Moreover, as the

vehicle got lighter it would accelerate beyond the 3g limit and this too had been a design ceiling for Shuttle engineers. The 'g' force limit had been imposed so that non-professional astronauts could fly aboard the Orbiter and carry out tasks, including experiments, without having had several years in training specifically for space flight.

The baseline – that all propulsion systems had to be controlled in thrust – was new to launch vehicles in the early 1970s. The early method used to align total thrust output with the remaining mass of the ascending vehicle was to build the launch vehicle with separate stages, throwing each away as it consumed its propellant. On the Saturn V, used to propel Apollo to the moon, each of the first two stages had five separate engines of fixed thrust. On most flights the centre engine of the five was shut down early to prevent over acceleration of the remaining mass as its propellant was depleted. With the Shuttle, however, the three main engines would have to fire from the launch pad almost into orbit, so the only option was to design them so they could be throttled down and then back up again when 'Max-q' had

ABOVE A deluge system protects the Orbiter from sound waves as water from 16 nozzles pours onto the flame deflectors beginning at T-6.6 seconds. At SRB ignition, 16 rain-birds pour additional water, with the total flow reaching 900,000 gal/min nine seconds after lift-off when shock waves are greatest. *(NASA)*

BELOW Astronauts inspect the powerheads on Orbiter main engines as part of their familiarisation and training routines. *(NASA)*

ABOVE Commander of STS-130 in February 2010, astronaut George D. Zamka stands alongside the RCC thermal insulation on the leading edge of *Endeavour*'s port wing. *(NASA)*

Space Shuttle Main Engine (SSME)

Consideration of the type of engine to use in the Shuttle began long before the design studies for the space plane proper got underway. In 1967 the US Air Force funded studies of advanced rocket propulsion systems that future launch vehicles might use. Rocketdyne was asked to look at a very advanced 'aerospike' engine, one in which the gases exhausted by the combustion process would flow down a central cone tapering to a point. The cone would provide the inner face of the expanding gases while the pressure of the atmosphere would provide the outer 'wall' against which the gases would flow.

This highly unusual concept originated with a need to solve an enduring problem with fixed nozzles: the shape of the nozzle dictates the efficiency of the gas flow shaping the exhaust; what is an ideal bell shape for low altitude within the earth's atmosphere is not the most efficient for operation in a vacuum. This is because the exhausted gases want to expand outward and only the dense atmosphere compresses the exhaust into a downward flow. As the rocket ascends, and the atmosphere gets thinner, the exhaust wants to bell out and take on the shape of the outer surface of an umbrella. Ideally, the bell should change shape and become much wider and flatter as the rocket ascends. If the pressure of the atmosphere itself is the outer wall of the nozzle, the exhaust cone will automatically adjust as the rocket leaves the atmosphere.

The aerospike concept was untried and highly advanced, but it was a viable way of providing one engine for two phases of flight: low down in the dense atmosphere and out into the vacuum of space. If the same engine was to be used for both booster and orbiter of the as yet undefined space plane, it was worth studying. But Pratt & Whitney (P&W) was asked to examine more efficient versions of conventional bell-nozzle types, less challenging technically. P&W opted for a high-pressure engine design operating on hydrogen and oxygen propellants, the most efficient for all practical purposes and one which would facilitate a more powerful and more compact engine.

been reached, but then down again as the Shuttle got much lighter.

The challenges with having to vary the thrust of both liquid and solid propellant rocket motors at specific times during ascent were immense. Coupled to the reusability of the Orbiter's main engines it made the Shuttle propulsion systems some of the most exotic ever built. As such they are a fascinating departure from the conventional rocket motors that had powered the Space Race since *Sputnik 1* in October 1957.

RIGHT For the Shuttle to perform as designed, it required the efficiency of a cryogenic propulsion system, engineering demonstrated by the two upper stages of the three-stage Saturn V moon rocket during the 1960s. *(NASA)*

P&W proposed a 250,000lb thrust engine designated XLR-129, but when the Phase A Shuttle feasibility studies came along in January 1969 the company offered a 500,000lb thrust engine with a fixed nozzle for the booster and a 510,000–590,000lb thrust engine with a two-position extendible nozzle for the Orbiter. As an alternative they put up a lighter engine with a thrust of only 415,000lb with a fixed nozzle, trading technological sophistication for simplicity and lower cost. When they began their own Phase A studies, the Shuttle bidders opted for this engine.

In February 1970 NASA invited Rocketdyne, P&W and Aerojet General to work up Phase B design concepts for the Space Shuttle Main Engine. Seemingly wrong-footed by focusing on the aerospike, which seemed to have been sidelined from the competition, Rocketdyne poured in a large amount of its own money to catch up. With their XLR-129, P&W seemed to be ahead and during the year they demonstrated a super-efficient hydrogen turbo pump and a very efficient engine producing 350,000lb of thrust. Aerojet General already had a lot of experience with big and powerful rocket motors, having designed a motor known as the M-1 with a thrust

ABOVE Preparations for the fifth flight, in January 1964, of a Saturn I carrying the cryogenic S-IV upper stage powered by six RL-10A3 liquid hydrogen/liquid oxygen rocket motors. *(NASA)*

LEFT The 90,000lb-thrust S–IV stage powered the last six of ten Saturn I launch vehicles. The routine handling of cryogenics was an important precursor to their adoption for the Shuttle Orbiter. *(NASA)*

RIGHT The next step up for cryogenic rocket stages was the S-II for the Saturn V, producing one-million-pounds of thrust. Built by North American Aviation it proved both troublesome and encouraging as eventual success brought confidence with high thrust cryogenics. *(NASA)*

BELOW The S-IVB third stage for the Saturn V was developed from the S-IV for the Saturn I and incorporated a single J-2 rocket engine of 200,000lb thrust, the same engine as that used on the S-II stage. *(NASA)*

of 1.5 million pounds. But NASA, plus the logic of a highly efficient engine requirement, argued for a staged combustion cycle with very high combustion chamber pressure.

Measuring performance

The efficiency of a rocket motor can be measured in a number of ways but the most useful is 'specific impulse', or Isp, which is the amount of thrust produced by one pound of propellant per second. If the ends of the equation are reversed, it can express the number of seconds one pound of propellant will produce one pound of thrust. So the term Isp is expressed in seconds – the higher the number the greater the efficiency of the engine or, put another way, the greater the amount of work it can do. Sheer power is a product of thrust measured in pounds force, but Isp gives a measure of how well the engine is doing at whatever thrust it is designed for. High chamber pressure is another way of improving the exhaust velocity and that provides a measure of the degree of acceleration in the reaction.

Rocket engines such as those used on the Saturn V moon rocket were of two kinds: those that used conventional fuel such as kerosene (a grade of paraffin, one variant of which in the UK is domestic heating oil), and those that used liquid hydrogen. Both fuels use liquid oxygen as the oxidiser essential for combustion. But the Isp of each is very different. Whereas kerosene/oxygen engines have an Isp number below 300 (seconds), a hydrogen/oxygen motor has an Isp above 400. If that is coupled to a very high combustion chamber pressure, the resulting reaction is very much greater again. If that can be achieved in a motor that weighs no more than a kerosene/oxygen engine then the performance is advanced even further.

The J-2 hydrogen/oxygen engines on the second and third stages of the Saturn V had a chamber pressure of 700lb/sq in, while the F-1 kerosene/oxygen engines on the first stage had a chamber pressure of up to 982lb/sq in. What Rocketdyne decided to do to get ahead of the competition was to build two 100-man teams under separate managers, one going for a significantly upgraded version of the J-2 engine, the second going out on a limb to build a high-pressure chamber, staged combustion design. All three engine bidders knew they had to go for very high chamber pressure and NASA appeared to be shaping the engine specification around a 3,000lb/sq in combustion chamber. To get the propellant into the combustion chamber operating at this pressure it would need another level of engine performance, which is where the staged combustion cycle came in.

Also known as a pre-burner cycle, staged combustion takes some of the propellant and combusts it in a pre-burner (a sort of miniature combustion chamber) from which the hot gases drive the turbines and the pumps that force the propellant into the main combustion chamber. Because the propellants passing through the pre-burners join the main flow of hydrogen/oxygen propellant, nothing is lost and for this reason it is sometimes referred to as a closed-cycle motor. But the engineering challenges are tough. Not only does the combustion chamber operate at 3,000lb/sq in, the temperature of the hot gases in the combustion chamber reach more than 6,000°F. Moreover, P&W's XLR-129 operated at this chamber pressure and it too was a staged combustion engine. New materials would be needed and new ways of operating rocket motors.

The winning bid

While all this was going on, NASA was required to upgrade the performance of the Shuttle to satisfy the military on whom it relied for support in getting it approved at the White House and in Congress (see Chapter 1). Also, during the closing months of 1970, NASA decided to reverse the decision to eliminate one of the three engines planned for the orbiter, which had been made to improve payload capability by 20 per cent due to reduced weight of the third engine. But a reversal of that decision was made so that there would be additional performance margin by having the third engine reinstated. It would make possible a better abort plan if the Orbiter had to return to the launch site in the event of trouble with the booster.

The really bad news came in January 1971, when NASA increased the baseline Orbiter payload capability to 65,000lb to a due-east orbit, a change mandated by the Air Force as a price for getting its approval to use the Shuttle for the majority of its missions. The thrust of each SSME would now have to go up from 415,000lb to 550,000lb sea-level thrust, and 632,000lb in a vacuum, and with the existing technical challenges already in place that was not easy. And there was an additional price to pay, too, for raising the thrust output and reinstating the third engine. With only limited space available at the rear of the Orbiter, the engines would have to be very close together

LEFT Shuttle engine development was a contemporary of flight operations with the Saturn V, yet the Shuttle engine was required to provide a maximum vacuum thrust of 512,000lb with a specific impulse of 453 seconds. *(Rocketdyne)*

which, added to the fact that they would have to be fitted inside the thrust structure in the aft fuselage, compromised heat dissipation.

Back at Rocketdyne this was good news – they were closer to a high-thrust engine than P&W – but they had yet to demonstrate a turbo pump and that went against the grain. Rocketdyne liked to build test and demonstration engines when submitting a design study or a bid and it badly needed to catch up with its prime competitor on that score. A novel way of getting ahead on

BELOW Proposed early in the Shuttle definition phase, the aerospike engine created an external burning surface with the pressure of the atmosphere forming the missing 'cone' shaping the expansion ratio. *(Rocketdyne)*

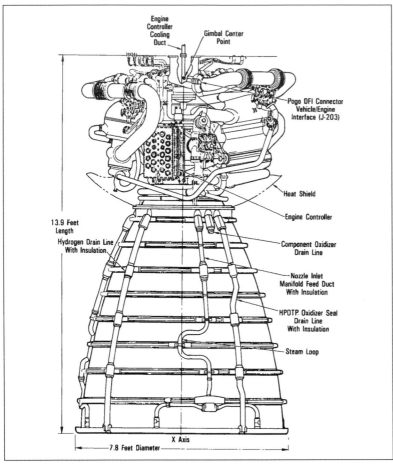

technical merit involved cooling. P&W had a combustion chamber cooled by labyrinthine passages for some hydrogen bled away from the main fuel supply and sprayed from the passages for transpiration cooling. But the drawback with this was the loss of a fraction of the fuel to cooling with no application to thrust output. This would lower the engine's specific impulse (Isp) by a few per cent and erode the value of a staged combustion cycle.

Rocketdyne had a completely novel solution to cooling the combustion chamber – making it out of a completely new material developed by one of its brilliant metallurgists, of whom it had a lot. The new metal was an alloy of copper and zirconium, with the thermal conductivity and twice the tensile strength of the former and the heat transfer properties of the latter. No hydrogen fuel would be needed to carry heat away from the combustion chamber walls because the material from which they would be made would be able to withstand the temperature. Rocketdyne was by this time a division of North American Rockwell and the new material was known as NARloy-Z. Under the management of the brilliant Floyd Bennett, a small test engine was fabricated from NARloy-Z

ABOVE The Space Shuttle Main Engine (SSME) was both complex in operation and compact in design, a veritable miracle of integration and optimisation. *(Rocketdyne)*

RIGHT The key to the operation of the SSME at high combustion pressures was the fuel and oxidiser preburners, which conditioned the propellants and utilising them for running the turbines in an environment where temperatures ranged from –423° to 6,000°F. *(Rocketdyne)*

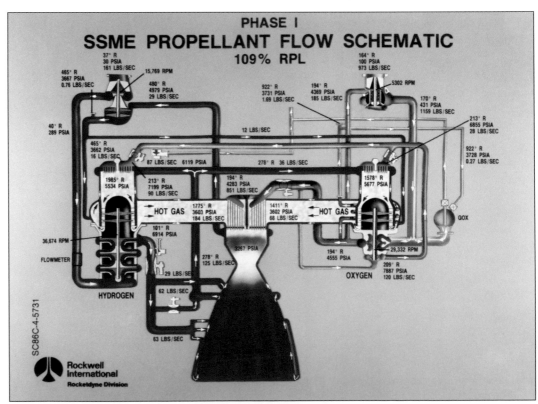

at chamber pressures of up to 5,000lb/sq in – and it worked!

Now it fell to the broad shoulders of Paul Castenholz, Rocketdyne's SSME project manager, to demonstrate a working engine before the bids went in on 21 April 1971. It is well to remember that at this intense time NASA still had accomplished only three of the six moon landings it would achieve, and that as far as the public was concerned the general pace of the space programme was still about Apollo and scientific expeditions to the moon. There was still the Skylab space station and a joint docking flight with the Russians out to the middle of the decade, but for industry it was make-or-break time. The Shuttle was the only big space project on the horizon and it was vital for space companies to get on board. Winning the SSME contract was absolutely essential to corporate futures, the work force transitioning from Apollo to the Shuttle, or to redundancy. Peak employment at Rocketdyne had plummeted from 20,000 in the mid-1960s to just 3,600 by 1970.

The customer drove the direction of the SSME design, leaving just enough room for individual design preferences, but the aerospike was out and the staged combustion engine was in. By early January 1970 Rocketdyne had a test engine ready to demonstrate all the salient features of their proposal. Under a test team led by Ted Benham, Rocketdyne engineers trucked a complete combustion chamber and powerhead with pre-burners to the company's Nevada Field Laboratory, about 20 miles from Reno. With no time to develop an electronic timed ignition device, Ted Benham positioned himself in a makeshift pillbox just 150ft from the test stand and when he saw the glow in the pre-burners he pressed a button manually to send the signal to start ignition of propellants in the combustion chamber.

This was intended merely to show the survivability of the NARloy-Z, a 3,000lb/sq in chamber pressure and good combustion gradient but the 0.35 second test fell just short of those goals. Nevertheless, on 12 February 1971, with thick snow on the ground, another test was set up and this time it demonstrated a chamber pressure of 3,172lb/sq in, with solid graph traces showing the NARLoy-Z holding

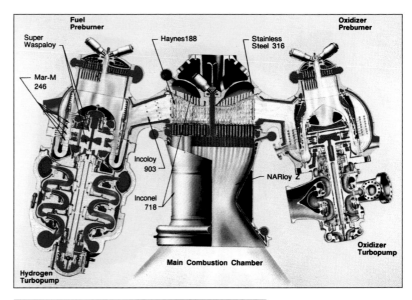

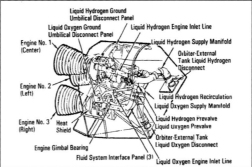

ABOVE The engine was designed for a chamber pressure of 3,000lb/sq in, more than three times that of existing rocket motors, which brought both temperature control problems and challenges for metallurgists. *(Rocketdyne)*

ABOVE Packaged tightly in the aft-fuselage section, the three main engines react against the thrust structure which transmits loads to the External Tank and Solid Rocket Booster assemblies. *(NASA)*

BELOW The powerhead carries the fuel and oxidiser preburners, an area much modified for the Mk II engines. The main injector assembly delivers propellant from the hot gas manifold to the combustion chamber. *(NASA)*

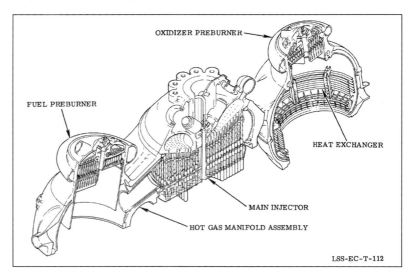

LSS-EC-T-112

RIGHT New technology demanded new techniques for manufacturing the critical components for the SSME, as with the electron beam welding process in use on this powerhead assembly. *(Rocketdyne)*

up and stable combustion. When Rocketdyne took their proposal to NASA in April 1971, Paul Castenholz gave the presentation of a lifetime and so impressed the customer that the company got the contract to build the SSME. P&W formally protested the decision, claiming at the time that it had been an unfair competition, but acknowledged later they were running bids for the F100 jet engine and could not have done both that and the SSME. Despite a legal battle, the protest was thrown out on 31 March 1972.

Defining the SSME

The contract for the SSME included detailed design, development, test, qualification and manufacture of 36 Space Shuttle Main

BELOW The combustion chamber must contain the 3,000lb/sq in pressure of engine operation for more than eight minutes of firing time on each flight, and is less than 14in in length from the injector plate to the exit throat. *(NASA)*

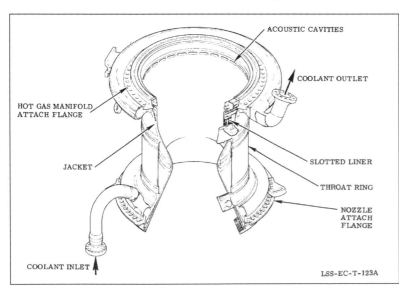

ACOUSTIC CAVITIES

COOLANT OUTLET

HOT GAS MANIFOLD ATTACH FLANGE

JACKET

SLOTTED LINER

THROAT RING

NOZZLE ATTACH FLANGE

COOLANT INLET

LSS-EC-T-123A

Engines, all to be completed and delivered by March 1978. But even as the ink was wet on the contract a big question arose over whether NASA could afford to build the Shuttle. Given a cost ceiling that was much lower than projected development costs, something had to give and in June 1971 the new NASA administrator James Fletcher gave formal consideration to what had always been rejected – phased development.

It was about this time that serious consideration was being given to replacing the piloted fly-back delta-wing booster with a more conventional rocket stage. A decision not to build the booster as originally conceived would mean the SSME would only be used on the Orbiter and maybe it, too, could be replaced by conventional rocker motors, albeit less capable and with less efficiency. Using a series of existing equipment would allow the Shuttle to come under the budget ceiling allowed to NASA, but them it would run the danger of never getting the more capable systems it really needed to do its job effectively. The fear was that once flying, an 'interim' Shuttle would be the only one NASA would ever get. It would also mean cancelling the SSME contract.

Fortunately, all the savings were made out of the booster and not the Orbiter or its propulsion systems, and the threat dissipated early in 1972 when President Nixon agreed to the proposed Shuttle. But now, with the fly-back booster replaced by a Solid Rocket Booster either side of the External Tank, the SSME would have to operate all the way from sea-level into space and the old problem of nozzle expansion ratios returned. Under the original plan the SSME would be applicable to both the booster and the Orbiter but the engines used in the manned booster would have had a much greater expansion ratio. This refers to the area of the nozzle at its base compared to the area of the combustion chamber exit throat where the gases expand into the nozzle.

Operating only in the very thin outer atmosphere and the vacuum of space, the SSMEs for the Orbiter would need a much greater expansion ratio. For the booster engines the expansion ratio was to have been 35:1, but for the Orbiter the ratio would have been 157:1, the latter facilitated by a nozzle, or skirt, extended after the Orbiter separated from the booster at altitude to greatly increase the area ratio and maximise the efficiency of the engine. In all other respects, SSMEs for the booster and the Orbiter would have been the same. However, since the elimination of a booster requiring the SSME, the expansion ratio would be set at 77.5:1 or less to optimise the performance of the engine across the broad altitude range. But there were some beneficial changes to dropping the fly-back booster. The original specification called for a capacity to throttle the engine down to 50 per cent of rated thrust, but with the new flight profile dictated by solid boosters that figure was reduced to 65 per cent.

How the SSME works

Propellants from the External Tank (ET) are delivered to the three SSMEs via feed lines with a diameter of 17in. On the launch pad the ET is pressurised with gaseous helium, which forces propellants to the engines via the liquid hydrogen feed line disconnect valve and in to the manifold. Here, equal volumes flow to each of the three main engines via a pre-valve which conditions the propellant for meeting the low pressure turbo pump. Each engine has the same components and the same cycle, which is as follows.

Hydrogen flows first to the Low Pressure Fuel Turbo Pump (LPFTP) inlet valve at a pressure of 30lb/sq in. The pump itself is about 18in x 24in and spins at 15,670rpm, increasing the fuel pressure to 303lb/sq in through a turbo-inducer. This prevents cavitation in the High Pressure Fuel Turbo Pump (HPFTP). The three-stage high pressure centrifugal pump is driven by a two-stage hot-gas turbine operating at 34,290rpm, which increases the pressure to almost 6,000lb/sq in. The HPFTP is 22in x 44in and is a generation away from the seven-stage axial pump utilised in the cryogenic J-2 engines used in the Saturn I and Saturn V programmes.

From the HPFTP, the fuel goes to the main fuel valve beyond which it diverges into three separate paths. Approximately 80 per cent of the hydrogen goes to the two pre-burners, half of which is first diverted to help cool the thrust chamber nozzle before it returns to join the

main flow to the pre-burners. The remaining 20 per cent of the hydrogen returns to help drive the low pressure fuel turbo pump. The liquid hydrogen is first routed to the main combustion chamber where it flows through 390 slots in the copper alloy walls. Converted into an ambient temperature gas, it is routed to the low pressure turbine where it is the energy source for driving the partial admission single-stage turbine for the low pressure pump.

Approximately 0.7lb/sec of this gas is routed back to another manifold connecting the other two engines to a return path into the ET, where it pressurises the main hydrogen tank, taking over that all-important responsibility from the helium used on the pad before the engines start up. The rest of it is used to cool the hot gas manifold (*see page 103*), together with the main

LEFT An engineer measures combustion chamber wall thickness using an ultrasonic micrometer at the Rocketdyne manufacturing plant, Canoga Park, California. *(Rocketdyne)*

BELOW The injector plate is a complex array of hollow posts spraying hydrogen and oxygen into the combustion chamber. The igniter is located in the centre of the plate assembly. *(NASA)*

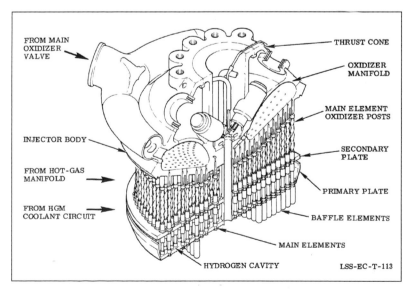

RIGHT Each injector post is carefully measured for size and inserted by hand into the face of the injector plate, where tolerances are fine and mistakes a precursor to catastrophe. *(Rocketdyne)*

injector baffles and faceplates, prior to it being consumed in the main combustion chamber.

The oxygen from the ET flows into each engine via a feed line, which separates into three 12in diameter flow lines, one to each engine. As with a fuel supply side, a pre-valve must be opened to allow the oxygen to move to the Low Pressure Oxidiser Turbo Pump (LPOTP). This is an axial flow pump spun up by a six-stage turbine, which raises the flow pressure from 100lb/sq in to 422lb/sq in,

RIGHT Looking up from the throat of the combustion chamber to the injector post array, where propellants are mixed and burned to produce thrust. *(NASA)*

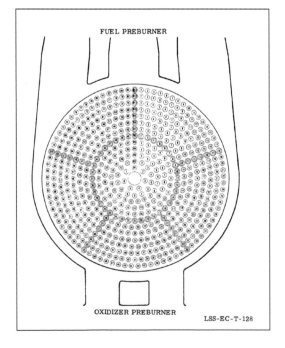

FUEL PREBURNER

OXIDIZER PREBURNER   LSS-EC-T-128

from where it is delivered to the High Pressure Oxidiser Turbo Pump (HPOTP). Approximately 18in square, the LPOTP runs at 5,150rpm and is flange-mounted on the propellant ducts. The boost in pressure to 422lb/sq in helps prevent cavitation in the high pressure oxidiser turbo pump.

The 24in x 36in HTOTP is essentially a main pump and a pre-burner pump consisting of two dual-inlet, single-stage, centrifugal pumps operating at 21,150rpm, raising the pressure from 422lb/sq in to 4,030lb/sq in. The discharged oxygen is diverted to several different pathways, the main flow going through the main oxidiser valve to the coaxial, main injector of the engine's combustion chamber. A small quantity (1.2lb/sec) is routed through to the engine-mounted heat exchanger from where it is returned to the ET. There, it is used to pressurise the main liquid oxygen tank in a similar process to the hydrogen side where it is used to pressurise the hydrogen tank. The balance of the flow is redirected back to a boost impeller on the same shaft to increase the pressure to around 6,940lb/sq in. This is sufficient to allow the throttle valves to control the liquid oxygen flow into the pre-burners.

Control of the thrust is achieved by throttling the oxidiser pre-burner side, and mixture control is achieved by using the fuel pre-burner side. A mixture ratio is necessary for optimising the balance between the ratio of fuel to oxidiser and can be used, much like an advance or retard in an internal combustion engine, to alter the richness of the combustion – another element in shaping the performance of rocket engines. For the SSME, this is an oxidiser/ fuel ratio of 6:1. The main throttle valves are controlled by the Main Engine Controller (MEC), a digital unit selected early on in preference to electro-mechanical or analogue systems, due to the finessed synchronisation required of the sequential steps needed for operation, control and throttle commands.

An equally important recirculation path taps power from the six-stage hydraulic turbine, which drives the low pressure oxygen turbo pump by bleeding 180lb/sec from the discharge side of the main impeller. After passing through the turbine, the gas is mixed with the discharging flow from the exit side of the turbo pump and returned to the High Pressure Oxygen Turbo Pump (HPOTP) inlet. Both pre-burners use a hydrogen-rich steam to drive the two turbines that drive the high pressure hydrogen and high pressure oxygen turbo pumps and these gases, too, make their way into the main combustion chamber to contribute towards thrust, hence specific impulse.

Because the high pressure oxygen turbine and pump are on the same shaft, a mixture of the fuel-rich gas in the turbine and the liquid oxygen in the pump could cause an explosion, so the two sections are separated by a purged cavity filled with helium gas. Sensors would detect a loss of helium pressure and immediately command an engine shutdown to prevent a disaster. Also for safety reasons, the low pressure oxygen and low pressure fuel turbines are assembled on opposing sides of the Orbiter thrust structure. Moreover, the ducts from the low pressure turbo pumps to the high pressure turbo pumps are attached by flexible bellows, while the two low pressure turbo pumps are fixed so that they gimbal along with the engine for directional control of the thrust axis. Liquid air is prevented from accumulating on the liquid hydrogen line between the low and high pressure fuel turbo pumps.

The hot gas manifold is the backbone to which the two pre-burners, the high pressure turbo pumps and the main combustion

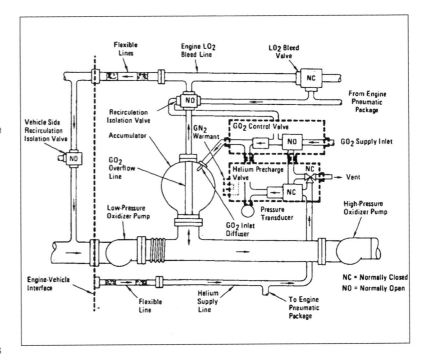

chamber are mounted. The liquid oxygen and liquid hydrogen flows to the pre-burners to be mixed, where they produce the hot gases which drive the high pressure turbo pumps. And it is through the manifold that the hot gases generated by the pre-burners flow to the main combustion chamber. The hot gases are introduced to the combustion chamber via the main injector, 17.7in in diameter, consisting of a barrel-shaped collection of 525 injector elements. Each element is a hollow cylinder

ABOVE To inhibit sympathetic frequency oscillations (POGO) in the fuel flow, gaseous oxygen is injected into the high pressure oxidiser inlet duct. *(NASA)*

LEFT After the SSME has been delivered to the Orbiter Processing Facility, a specially built fork lift with lifting fixtures and restraining lugs manoeuvres the unit toward the No.1 (top-centre) position. *(NASA)*

RIGHT The No.3 SSME
is gently manipulated
into the thrust
structure in the Orbiter
aft-fuselage, with
several technicians
inside guiding the fork
lift driver. (NASA)

through which the hot gases flow to be injected into the main body of the combustion chamber. Concentric rings of injector elements ensure a spray of propellants for final combustion under a pressure of up to 3,277lb/sq in.

To aid stable combustion and to prevent vortices spiralling up the sidewalls, 75 baffle elements, each with a length of 2in, are arranged in five groups of 15 arranged radially at 72° intervals around the circumference of the main injector plate. Various tests are conducted to ensure that combustion instability cannot take place but that, if some unexpected transient gas flow should begin to create one, the baffles will damp it out and restore combustion stability.

The injector exit diameter is 17.7in, but the main combustion chamber has a throat area of 83.4sq in with a contraction ratio of 2.96:1 and an expansion ratio of 5:1. The length of the combustion chamber from the injector elements to the throat is 14in. Ignition is provided by a dual-redundant igniter located in the centre of the injector plate. They are turned off three seconds after ignition as the combustion process is self-sustaining once the engine is up and running.

From the throat of the combustion chamber to the bottom of the main expansion skirt the nozzle has a length of 10ft 1in, and with an internal exit diameter of 7ft 6.9in, it has an exit area ratio of 77.5:1. The nozzle was upgraded later in the Shuttle programme to have a length of 9ft 5in, creating an expansion ratio of 69:1. The nozzle has 1,080 coolant passages fed at three inlets with gaseous hydrogen which, although never 'cold', is at a much lower temperature than the wall of the expansion skirt.

One vital part of the engine is the pogo-suppression system. Pogo is the effect created by low-frequency flow oscillations which, when coupled to the natural frequency of the vehicle, can cause dangerous vertical oscillations, much like a pogo-stick, from which the name has been borrowed. Pogo, which has been a problem with rocket stages since the first large liquid propellant motors, is solvable through the use of a gas accumulator which, on the SSME, is attached to each high pressure oxidiser inlet duct. The gas disrupts transmission of these low frequency oscillations so that it never builds to become a dangerous phenomenon and comprises a 0.6cu ft accumulator containing a standpipe, gaseous oxygen valve and two recirculation isolation valves. Charged with helium 2.4sec after engine start, and until the engine heat exchanger can provide gaseous oxygen, the accumulator is chilled with liquid oxygen during engine chill-down.

The installed position of the three main

engines is canted up and out from the longitudinal axis of the Orbiter. Taking a line straight through the centre of the fuselage, the No. 1 (top) engine is canted upward 16°. The No. 2 (left looking forward) and No. 3 engines are canted upward 10° and outward 3.5° from the centreline, so that they are pointing rearward 7° from each other. To control the Shuttle in flight it is necessary to gimbal the engines to align the thrust axis with the centre of mass or, to offset that alignment to allow it to roll or pitch under full control on to its assigned trajectory. To do that two hydraulic actuators are provided for each engine operated by two of the three Orbiter hydraulic systems. The pitch actuator can move the engine 10.5° up or down and the yaw actuator a maximum 8.5° up or down. The engine gimbal actuators can move the engines at a maximum rate of 20°/sec.

Making it better

Development and testing of the SSME was a long and protracted process. The catalogue of failures and modifications, as well as numerous changes to the way the SSME was built, would fill volumes of books. Suffice it to say that the challenges were extraordinary and the development of the SSME has been the only major rocket engine programme in the

United States since the 1970s. Everyone knew that it would have to go through a sustained programme of upgrades and improvements. Because the Shuttle was never an operational vehicle every mission added to the building blocks of knowledge about this unique machine, with flight experience continuously fed back into a series of redesigns and improvements.

When Shuttle flights began in the early 1980s, the amount of time needed to maintain the engines and the failures on test of engines supposedly demonstrating higher levels of performance, brought calls for a major effort to upgrade the engine design. Three phased development programmes were planned: to reduce the between-flight maintenance required; to upgrade power output to 109 per cent of rated thrust; and to provide long term performance margin for the US Air Force flights (which, then, were planned for flight from Vandenberg Air Force Base in California).

In truth, there were several issues with the operational history of the SSME on Shuttle flights. After ignition of the No.3 SSME, four seconds prior to launch of STS-41D on 26 June 1984, a sensor detected loss of redundant control on the main fuel valve, which promptly shut down that engine two seconds after ignition of the No.2 engine, and immediately

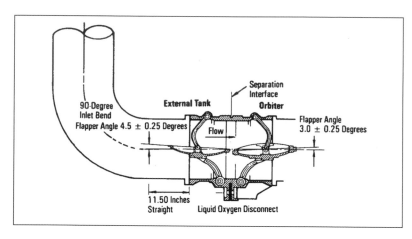

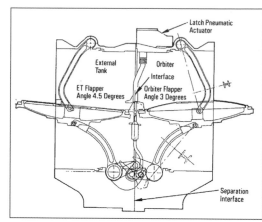

before ignition of the third engine. This was the first launch pad abort but on 12 July 1985, it happened again when the No.2 SSME was shut down when the hydrogen coolant valve failed to close at T-3 seconds. Recycled to 29 July, just 5min 45sec into the flight the No.1 SSME was shut down by the general purpose computer after detecting excessive temperatures once the engines had been throttled back from 104 per cent to 65 per cent.

This would not be a major problem, with the other two engines being throttled up to 91 per cent and burning 70 seconds longer to compensate for the slower ascent, but when another sensor on a second engine indicated the same thing, quick thinking on the part of a ground controller advised the crew to bypass

the sensor and keep the engine running. The telemetry data on the ground clearly indicated that it was a sensor failure and not a realistic condition. Had the second engine shut down the Shuttle would have had insufficient energy from the remaining engine to make it into orbit, necessitating an abort to an emergency landing site on the other side of the Atlantic Ocean at Zaragoza in Spain.

After the *Challenger* was destroyed during ascent on 28 January 1986, a thorough review resulted in the cancellation of the 109 per cent thrust SSME upgrade development and the one to build in long-term performance margins. By this time the Air Force had cancelled plans to fly the Shuttle from Vandenberg Air Force Base and turned its back on the Shuttle as a

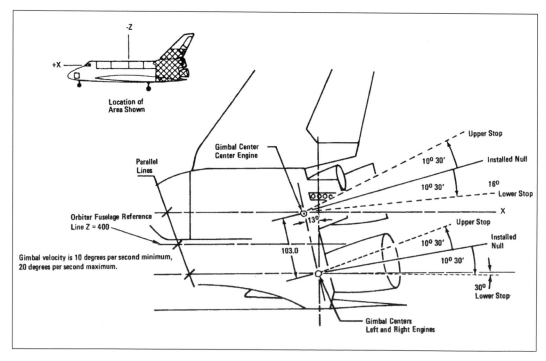

means of gaining access to space, preferring to fund further development of conventional, expendable, rockets that were clearly cheaper to launch. Nevertheless, a steady progression of improvements to make the SSME more reliable and widen the safety margins were instituted over the next 25 years.

Ultimately, in the wake of the report into the *Challenger* disaster, although the SSME was exonerated of any role in causing the event, there were serious concerns about safety and several engine components were earmarked for upgrade. These included the high pressure fuel turbo pump, the high pressure oxidiser turbo pump, the powerhead, the heat exchanger and the main combustion chamber. In August 1986, NASA issued P&W with a contract for 44 alternate high pressure oxygen turbo pumps to be delivered to Rocketdyne under a Block I programme. A second, Block II, development would follow with an advanced high pressure fuel turbo pump, together with a large diameter main combustion chamber – one development effort Rocketdyne had recommended to NASA in 1980.

With 50 per cent fewer rotating parts and with a unique casting process, negating the need for all but six of the 300 welded parts requiring meticulous inspections, the new P&W turbo pumps were more reliable and potentially less prone to failure. With ceramic bearings up to 30 per cent harder and almost half the weight of the Rocketdyne bearings, it was hoped that the bearings would last longer than the two flights previously experienced with the original designs. The new pump operates at 23,700rpm and generates 25,850shp with a discharge pressure of 7,250lb/sq in. Development test firings started in February 1990 but serious delays to the upgrade programme were induced by budget constraints and the inevitable problems encountered during testing.

Changes defined as Block I improvements also included a new two-duct powerhead instead of the original five-duct design – three ducts to collect gases from the fuel turbine; and two on the other side of the engine collecting gases from the oxidiser turbine. This helped improve gas flow, reducing both pressure and dynamic loads and eliminating 80 welds and 52 components. Improvements were also made to the engine's heat exchanger, which is a piece of coiled stainless steel tubing, 41ft long, transporting liquid oxygen to pass through the hot gas manifold and heated by the exhaust from the high pressure oxidiser turbo pump. The new continuous single-coil heat exchanger eliminates all welds and is more efficient, as well as costing a lot less to manufacture.

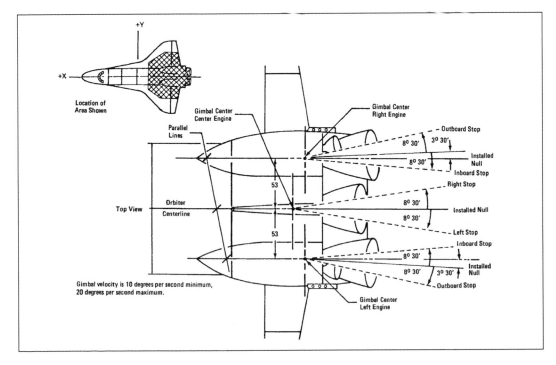

LEFT Whereas pitch movement provides a traverse of up to 21°, as shown on the opposite page, yaw motion is limited to 17°, divided in equal deflections either side of the null position. *(NASA)*

Another change recommended by Rocketdyne in 1980 was the development of a large throat main combustion chamber, but while this was thought to improve specific impulse, albeit only slightly, it was also believed to be prone to combustion instability. Whereas the old design had many separate sections welded together, the new chamber has many more castings, eliminating more than 50 welds while providing an 11 per cent increase in throat area. In turn, this allows for lower turbine temperatures, improved cooling, a reduction in maintenance and longer life. But it gives an extra 1.4 seconds of Isp and chops power demand at the high pressure fuel turbo pump by 3,000hp.

Making it safer

These and many other improvements added greatly to projected reliability rates. In the 24 Shuttle flights flown between 12 April 1981, and 12 January 1986, estimates as to potential Shuttle loss rates due to a catastrophic disaster between launch and landing ran wildly between mathematical calculations based largely on uncertainties. From the outset, NASA relied on calculations based on qualitative methods for assessment of Shuttle risk instead of quantitative methods, as recommended by the US National Research Council in all high-risk activities. Qualitative assessment is based on numerical values applied by the reasoning of individual engineers and is therefore subjective.

What NASA failed to do right up to the loss of *Columbia* in 2003 was to import a quantitative analysis, where probabilistic risk assessment is used as a systematic (objective) method for evaluating the possibility that an event will occur and predicting the consequences. The author was heavily involved in risk assessment of the Shuttle transportation system up to the *Challenger* accident of 28 January 1986, and for several years after the Shuttle resumed flight operations on 29 September 1988. The methods used to calculate the likelihood of total loss, including the lives of the crew, were based on spurious parameters which took little account of the realities of what was, to the end of all Shuttle flights in 2011, an experimental vehicle that everyone wanted to think was an operational system.

Merely launching the Shuttle requires 1.2 million separate procedures and NASA has identified more than 5,000 critical components, in which the failure of any one would result in a total loss of the vehicle. Only after the loss of *Columbia* on 1 February 2003 did a measure of reality sink in, although by then major efforts had been underway for several years to improve the safety and reliability of the system. Considerable attention focused on the probability of a catastrophic engine failure which, given the nature of the Shuttle configuration, could be expected to result in the total loss of the Shuttle and its crew *(see Chapter 5 for abort criteria)*.

Prior to the loss of *Challenger* in 1986 the estimated loss rate for the Shuttle throughout its entire mission was anything between 1 in 1,000 to 1 in 500 flights. After *Challenger* the figure came down to 1 in 78. In dissecting the mission into segments and then into systems, the probability of the failure of a main engine causing catastrophic loss during ascent the figure was 1 in 278. With Block I SSME improvements that figure went up to 1 in 335. With Block II upgrades the predicted loss rate increased to 1 in 483 flights. But these figures relate only to the probability of an SSME failure causing total loss of the Shuttle during the ascent to orbit. For all mission phases, total probability of a failure at some point in a Shuttle mission is still greater than 1 in 150.

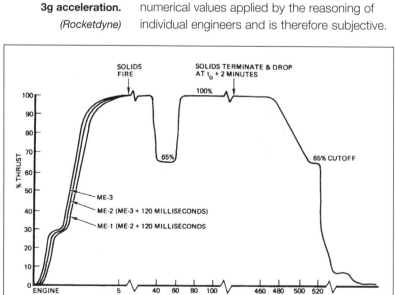

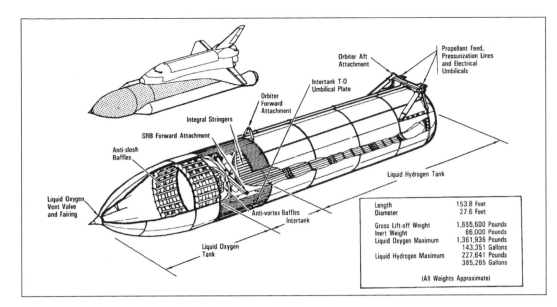

Orbiter Aft Attachment

Propellant Feed, Pressurization Lines and Electrical Umbilicals

Intertank T-0 Umbilical Plate

Orbiter Forward Attachment

Integral Stringers

SRB Forward Attachment

Anti-slosh Baffles

Liquid Oxygen Vent Valve and Fairing

Anti-vortex Baffles Intertank

Liquid Oxygen Tank

Liquid Hydrogen Tank

Length	153.8 Feet
Diameter	27.6 Feet
Gross Lift-off Weight	1,655,600 Pounds
Inert Weight	66,000 Pounds
Liquid Oxygen Maximum	1,361,936 Pounds
	143,351 Gallons
Liquid Hydrogen Maximum	227,641 Pounds
	385,265 Gallons

(All Weights Approximate)

LEFT The layout of the External Tank is basically a conventional rocket stage, but with the rocket motors displaced to the rear of the orbiter. The cryogenic hydrogen is stored below the liquid oxygen. *(NASA)*

RIGHT Built by Martin Marietta, the first External Tank assembly is rolled out to public view on June 29, 1979, but it will be more than 21 months before it is launched on the first Shuttle flight. *(Martin Marietta)*

External Tank (ET)

When the Shuttle was originally conceived in the mid-1960s, all the propellant (liquid hydrogen and liquid oxygen) was to have been incorporated inside the structure of the Orbiter, but as a means of saving development costs the propellant tanks were removed and placed within a single, cylindrical structure known as the External Tank (ET). The ET would supply propellant for the three Space Shuttle Main Engines all the way up into orbit and, to prevent the ET from becoming a hazard in space, a small retro-rocket would de-orbit the tank into a remote ocean area where most of it would burn up in the atmosphere.

But why carry the ET all the way to orbit when the SSMEs could shut down just before achieving orbital velocity, allowing the ET to separate and fall back down into the atmosphere due to gravity, removing the need for a retro-rocket in the nose of the ET? The External Tank would separate marginally short of achieving orbit leaving the Orbital Manoeuvring System (OMS) engines at the rear of the Orbiter to add the final push into orbit. This simplification would bequeath to the OMS engines responsibility for nudging the Shuttle into orbit.

Although the ET appears to be a simple

LEFT The first two tanks were sprayed with an inflammable latex coating, but later tanks dispensed with that, leaving a white foam insulation that turns to yellow in sunlight.

(Martin Marietta)

BELOW Although never carried to orbit inside the Shuttle, General Dynamics designed cryogenic hydrogen/oxygen fuelled Centaur upper stages for boosting heavy spacecraft to the planets. In the top background is a widebody Centaur for a Titan expendable launch vehicle.
(General Dynamics)

piece of engineering, being a cylindrical structure comprised of a forward (liquid oxygen) tank and an aft (liquid hydrogen) tank, with an inter-tank structure between the two, it is in reality a complex structure responsible for transmitting all loads between the elements attached to it. In effect, the ET is the structural backbone to which the two Solid Rocket Boosters and the Orbiter are attached. With the primary contract for the Orbiter given to North American Rockwell on 26 July 1972, on 2 April 1973 NASA released a request for bids to build the External Tank to Boeing, McDonnell Douglas and Martin Marietta.

All three had experience in building launch vehicles or rocket stages: Boeing the giant S-IC first stage of the Saturn V, McDonnell Douglas the S-IV and S-IVB stages of the Saturn I and Saturn IB, and Martin Marietta the Titan 1 and Titan 2 ballistic missiles, plus several evolving adaptations of these missiles as space launch vehicles. Because of its contract to build the Orbiter, Rockwell by itself was prohibited from competing but they teamed with Chrysler, responsible for building Saturn I stages, in sending in a fourth bid.

When they came in, all four were viable to varying degrees but Martin Marietta had a unique selling point. Because the Titan 2, adapted as a satellite launcher, was in essence a rocket stage flanked by two solid rocket boosters it was the only company with experience of a similar concept to the one chosen for the Shuttle. Martin was unbelievably low on cost projection but it got the work in a contract dated 16 August 1973. Potentially, with NASA then expecting to fly up to 60 Shuttle missions a year, this was good business as each flight would need a new tank, the only element of the Shuttle completely expendable.

ET configuration

The External Tank is very big. At 153.8ft long and with a diameter of 27.6ft it consists of a forward tank containing liquid oxygen and an aft tank holding liquid hydrogen. For reasons of space these propellants are stored as liquids, their volumes as gases being far in excess of what would be physically manageable. But liquid oxygen boils at –297°F and hydrogen boils at –423°F so to contain these fluids within

RIGHT Specially developed techniques for connecting External Tank panels include friction stir welding, which uses frictional heating combined with forging pressure to produce strong bonds free from defects. *(Martin Marietta)*

ABOVE The completed oxygen tank is lifted into position at the assembly plant in the Michoud Assembly Facility. *(Martin Marietta)*

RIGHT The cylindrical hydrogen tank has a diameter of 27.5ft and utilises the friction stir welding technique. The first tank to adopt this innovative technique was ET-134. *(Martin Marietta)*

BELOW The first sections to be fabricated using the friction stir weld process were longitudinal barrel sections on both the oxygen and hydrogen tanks. The author helped bring this technology out into general industry. *(Martin Marietta)*

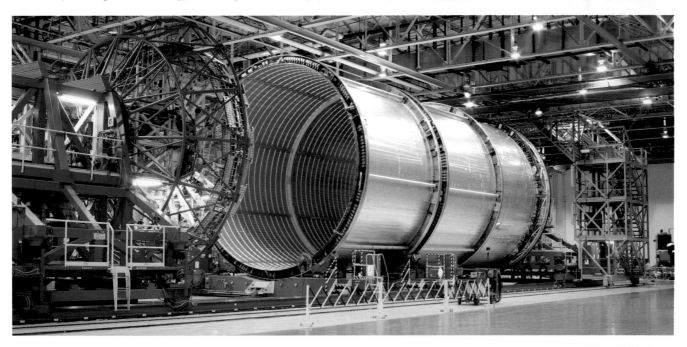

manageable proportions, the External Tank is a giant thermos flask designed to keep the cold in and the heat out.

Fluids can be held in either the gaseous or the liquid state, depending upon the temperature, and it is fascinating to reflect that aviation took off into a fluid (oxygen and nitrogen) encapsulating the earth as a gas, whereas high performance rocketry gets off the ground on fluids (oxygen and hydrogen) as liquids! Because of their super-cold properties, they are known as cryogenic propellants.

In the late 1940s and early 1950s there had been much research into cryogenic propellants, one area of expertise in which US rocket engineers excelled compared to their Russian counterparts. The first cryogenic stage to fly was the Centaur, which evolved out of a pool of highly secret technology developed by the Air Force for a top-secret hydrogen fuelled spy-plane known as Project Suntan. Within three years, anticipating an age of hydrogen-burning aircraft, the capacity for the US to produce

LEFT The inter-tank section separating hydrogen and oxygen tanks is stiffened by hat-section stringers. (Martin Marietta)

BELOW In this view, peering though an access door into the inter-tank section, the top of the hydrogen tank almost meets the end dome of the upper oxygen tank. (NASA)

liquid hydrogen increased from 500lb a day in 1956 to 68,000lb a day by 1959.

With significant advantages known from basic chemistry, the attraction of liquid hydrogen/liquid oxygen rocket motors was too great to pass up and development of the world's first cryogenic rocket motors began in the late 1950s. Powered by two 15,000lb thrust RL-10 cryogenic engines built by Pratt & Whitney, the Centaur upper stage for the Atlas launch vehicle was first flown successfully on 27 November 1963. This was followed on 29 January 1964 by the first Saturn I launch vehicle carrying the first S-IV hydrogen/oxygen upper stage, powered by six RL-10 rocket motors.

Precursor to the 200,000lb thrust Rocketdyne J-2 cryogenic engine used in the Saturn IB and the upper two stages of the Saturn V moon rocket, the RL-10 and the Centaur became the workhorses for satellite launches from the mid-1960s, while the J-2 provided much experience for Rocketdyne to work on the Space Shuttle Main Engine. Moreover, the experience with handling, moving and operating cryogenic propellants came in very handy when NASA stipulated a cryogenic propulsion system for the Shuttle. The essential technology was in place and much as the Shuttle benefited from unique management structures adopted from the Apollo programme, cryogenic rocketry gave industry a head start in this advanced form of propulsion.

The forward section of the External Tank is where the liquid oxygen (LOX) is stored, a vessel with a conical forward section and a domical aft section, 49ft 5in long and a diameter of 27ft 7in, with an empty weight of 12,000lb. It has an internal volume of 19,182cu ft and is capable of holding 1,391,936lb of liquid oxygen, equivalent to 143,351 US gal. The first tanks were fabricated from 2219 aluminium, but later tanks were made from 2915 aluminium-lithium for weight saving reasons. The LOX tank is usually pressurised at 20–22lb/sq in and contains anti-slosh baffles to prevent the liquid oxygen creating wave-like motion and throwing the vehicle off course.

LEFT The nose cap of the External Tank (ET) where gaseous oxygen in vented to prevent over-pressurisation prior to lift-off. Originally, this was the location for a de-orbit rocket when it was anticipated the Shuttle would reach orbit on the three main engines alone. *(NASA)*

BELOW The liquid oxygen feed line emerges from the inter-tank section to run down the exterior of the hydrogen tank. *(NASA)*

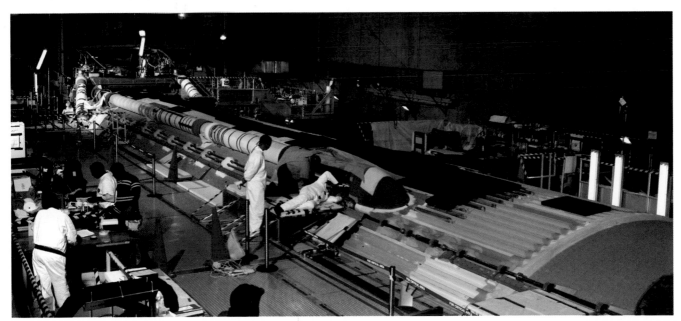

RIGHT The Orbiter is attached to the ET by means of a bipod on the inter-tank section and two fixtures at the rear of the hydrogen tank, where propellant enters the aft-fuselage and is delivered to the main engines. *(NASA)*

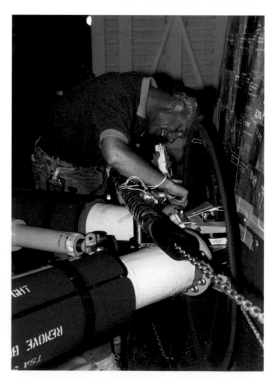

The tank delivers LOX through a 17in diameter feed line that carries LOX down through the inter-tank structure and then to the exterior of the ET, where it travels down the side of the liquid hydrogen tank. The feed turns up into the thrust structure of the Orbiter at the right-hand ET/Orbiter interconnect, where it is delivered to the main engines at a flow rate of 2,787lb/min (17,592gal/min). The LOX tank contains a vent valve which opens after separation to provide a tumbling motion as residual liquid oxygen flows out into space, acting like a rocket thruster and creating a reaction. By doing this the ET is tumbling when

it slices back into the atmosphere and is more likely to be destroyed.

The inter-tank section is 22ft 6in long with a diameter of 27ft 7in and weighs 12,100lb, constructed of 2219 aluminium alloy or 2915 aluminium-lithium in later, lightweight, structures. It is of semi-monocoque construction with flanges at top and bottom for joining to the adjacent propellant tanks above and below. The inter-tank also houses the thrust beam, serving as the forward attachment points for the two Solid Rocket Boosters, and it incorporates additional fittings for transmitting loads between the upper and lower propellant tanks. The inter-tank is vented in flight and so becomes a vacuum in space.

The liquid hydrogen (LH_2) tank is the largest structure in the ET assembly, 96ft 8in long and 27ft 7in in diameter, weighing 29,000lb empty. Early tanks were built up from fusion-welded 2219 aluminium and changed to 2915 aluminium-lithium in later tanks, built to operate at a pressure of 32–24lb/sq in. The structure is assembled from four separate barrel sections, five major ring frames between each and at opposing ends, with forward and aft ellipsoidal end domes. There are anti-vortex baffles inside and a syphon outlet to move the LH_2 to the thrust structure of the Orbiter via a 17in

RIGHT The propellant feed disconnect and flapper valve assembly between the ET and the Orbiter. A close-out door seals the undersurface with thermal protection tiles. *(NASA)*

FAR RIGHT The bipod connection where the Orbiter is attached to the forward thrust attachment on the ET is checked using strain gauges. *(NASA)*

RIGHT Spray-on foam insulation helps maintain internal cryogenic temperatures within the lithium-aluminium tank. *(NASA)*

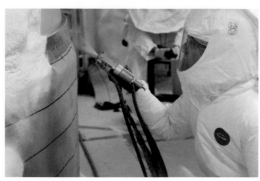

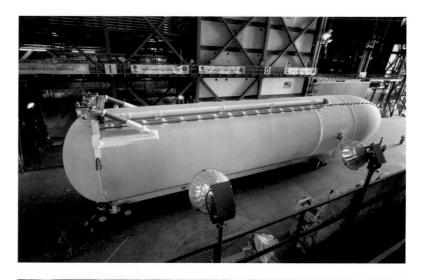

diameter feed line to the left side of the ET/
Orbiter interconnect.

Because hydrogen has such low mass, there
is no need for anti-slosh baffles since wave
motion would induce little reaction on the tank
walls. It can hold 237,641lb of liquid hydrogen,
or 385,265 US gal within its internal volume of
51,737cu ft. The LH_2 tank also supports the
attachment points for the Orbiter. At the forward
end of the tank is the forward Orbiter bipod
attachment strut, with the aft attachment ball
fittings at the base of the tank where the ET/
Orbiter disconnects are located. This is where
the propellants are fed into the thrust structure
of the orbiter through snap-shut disconnect
doors on the underside of the aft-fuselage.

Four depletion sensors are fitted to the
bottom of each tank, but engine cut-off is
usually managed as a function of velocity. If,
however, depletion of either propellant is sensed
the engines are commanded off and the position
of the LOX sensors allows the maximum amount
of oxygen to reach the engines to prevent
cavitation in the oxidiser pumps where they run
dry. Propellant loading provides 1,100lb of LH_2
above that required for combustion so that at
engine cut-off the propellants are fuel-rich to
prevent damage that would be caused by an
oxygen-rich condition. The ET carries electrical
and telemetry data channels to the SRBs and
the Orbiter so that sensors can communicate
with the general purpose computers.

To contain the internal temperatures and to
insulate the ET from thermal leaks, the entire
outer surface is coated with a CPR-488 spray-
on foam insulation (SOFI) with a charring ablator
to prevent high heating during ascent caused
by friction with the atmosphere. The tanks used
on the first two Shuttle flights had a white fire-

RIGHT When Shuttle flights were frequent, ETs
were hoisted high into the 525ft tall VAB, where
they could be seen hanging high in the 'loft'
looking like cocoons. (NASA)

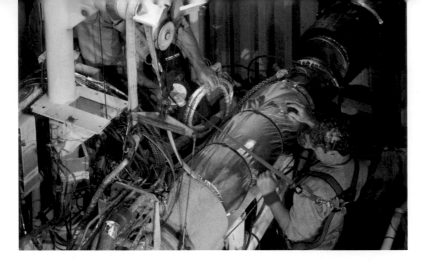

retardant latex coating but this was removed as a weight-saving measure, reducing the weight of the insulation by 595lb. Nevertheless, the SOFI coating still weighs 4,823lb.

The relative placing of the Orbiter on the External Tank, and the potential danger of sections of the insulation falling off during ascent and damaging the Orbiter's brittle carbon-carbon wing leading edge heat protection was disregarded. Following the launch of *Columbia* in January 2003, a piece of the ET SOFI came off and struck the Orbiter's wing, gouging a piece of the wing leading edge which, on re-entry, allowed heat from friction with the atmosphere to enter the wing and melt the aluminium structure, resulting in the loss of the spacecraft and its crew of seven.

Slimming down

A Light Weight Tank (LWT) was introduced to the Shuttle beginning with the sixth flight, the first flight of *Challenger* on 4 April 1983. This was achieved through reducing the number of stringers in the inter-tank section and milling down some of the skin panels. The anti-slosh baffles were also redesigned, saving more weight, and eliminating a recirculation line. All of the 5A1-2.5 titanium fittings were changed to the more commonly used alloy 6A1-4V and all 7075-T73 aluminium was changed to 7050-T73 aluminium, which allowed an increase in strength but a reduction in weight. The new lightweight tanks had an empty weight of 66,800lb, about 10,300lb lighter than those flown on the first five flights.

When *Atlantis* was launched on 2 December 1988, damage to some of the Orbiter's tiles was caused by insulation breaking free from the nose cone of the External Tank. In June 1989 a redesign of the nose cone was authorised, replacing the original design, which had incorporated several metal sheets and more than 1,100 fastenings in a structure necessitating foam insulation. The new one-piece cone was fabricated from a polymeric-matrix composite, with high heat resistance capable of withstanding 900°F. Following a substantial period of testing it was applied first to ET-81 flown on 19 January 1996, and used on subsequent flights.

The final weight reduction programme produced the super lightweight tank (SLWT) which joined the Shuttle fleet with the launch of *Discovery* on 2 June 1998. Martin Marietta began work on the SLWT in 1991 and introduced new metals and fabrication techniques to reduce weight, lower manufacturing costs and increase the operability of the tank design. The Martin Marietta laboratories developed a new aluminium-lithium alloy (2195) which proved to be 40 per cent stronger and 10 per cent less dense than the standard 2219 alloy previous tanks had been made from. It consisted of 94.2 per cent aluminium, 4 per cent copper, 0.4 per cent magnesium and 1 per cent lithium, the balance made up from sundry chemicals, and proved effective, if costly, from which to fabricate External Tank assemblies.

Unfortunately, new aluminium-lithium alloy is prone to contamination in the welding process so Martin Marietta designed a new chamber, which would purge the area being worked on to eliminate that possibility. The SLWT programme

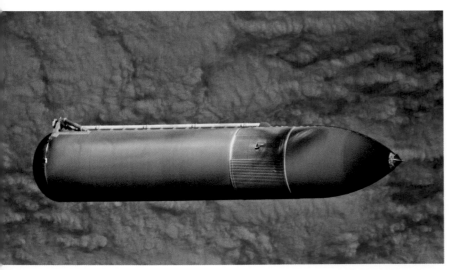

did more than change the materials. Although the new al-li reduces overall tank weight by 4,500lb, a further 2,500lb is shaved from replacing the LH_2 tank's waffle-pattern grid design with T-stiffeners. And a further 500lb is saved by utilising a SOFI contour map, which allows the foam insulation to be applied in varying thickness depending on the specific area. Overall, this reduces weight by a further 7,500lb. The LWT and SLWT programmes have reduced the total dry weight of the External Tank by 17,800lb, or 23 per cent.

Solid Rocket Booster (SRB)

When the suggestion was made that solid propellant boosters could be used to get the Shuttle off the pad and towards orbit, there were many questions that brought concern. Unlike liquid propellant rocket motors, solids are giant fireworks, packed with a rubbery substance combining fuel and oxidiser. They are usually of fixed thrust and provide little opportunity to shut them down if something goes wrong. They can be built to have great thrust but they are the least efficient of all forms of rocket propulsion, the specific impulse of even the most favourable solid propellant combinations being much less than that of the worst liquid propellant motor. Once ignited they will burn until their propellant is depleted and they are unlikely contenders for reusability.

On the other hand, they are less complicated than liquid propellant rockets, are cheaper to develop, cheaper to build and are more robust in both handling and operation. Because of their negative properties they had never been considered for carrying humans into space, but when cost was paramount over efficiency – as it was with the crucial Shuttle decisions in the second half of 1971 – they were an accountant's choice. Not that there had not been a lot of development already with solid propellant rockets.

Although the first ballistic missiles used liquid propellant motors, solid propellant rockets had been around in military use for centuries. It had been with William Congreve's solid propellant rockets that the first White House had been burned down during the War of 1812, an event

enshrined in the national anthem of the United States. By the Second World War small-scale solid propellant rockets were in wide use and by the 1950s they were being developed for long range applications. The US Navy chose solids for its Polaris submarine-launched ballistic missiles and by the early 1960s the Minuteman ICBM was America's land-based nuclear deterrent.

Aerojet was one of the first to exploit the technology of segmented solids, making them easier to move around and bolt together for flight at the launch site, freeing them from the limitations of rail trucks and road cars. The military got interested and in several test projects into the early 1960s, giant solids were being built for analysis with a view to their use as cheap space launchers. The power of the solids was proportional to their size and manufacturers began experimenting with solids of increasing diameter. In May 1964 Lockheed tested a 156in diameter solid 70ft long and demonstrated a thrust of 949,000lb for 1min 48sec. They emptied the residue and refilled it with more propellant then fired it again. That time it generated a record thrust of 1.1 million pounds. Several more tests took place, the big solid proving it would work.

This was followed by a 260in diameter solids programme, taken up by NASA's Lewis Research Center with bigger cases of even greater thrust, finally resulting in a test firing conducted by Aerojet on 17 June 1967, which produced a thrust of 5.884 million lb of thrust for 77sec, clearly a record for a single rocket motor. But there were no uses for such a

BELOW The Solid Rocket Boosters (SRBs) comprise Solid Rocket Motor (SRM) segments joined into four separate 'factory' segments, delivered by rail and stacked in the VAB at what are known as field joints. (NASA)

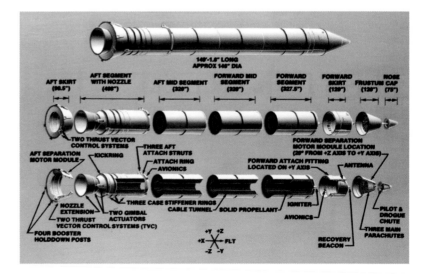

development cost that would bring the Shuttle under the funding bar set by the White House and by Congress. Following 18 months of study using contractors to examine options, bids to build the Solid Rocket Boosters were requested by NASA on 16 July 1973. Many engineers had wanted to build in a thrust-termination system so that the Orbiter could be separated from boosters running amok, but that went out to save 8,000lb in weight.

The requirement was relatively simple: provide a 156in diameter SRB that could produce a thrust of around 3 million pounds for little over two minutes, separate from the External Tank and be recovered using parachutes to lower the massive booster to the Atlantic Ocean off Cape Canaveral. Piercing into the water base down, it would float vertically, one-third above the water, until divers could go down and plug the open nozzle. Pumping in air would cause the booster to pitch over and float on the surface, where a recovery boat would tow it to Port Canaveral so it could be cleaned out and prepared for another launch. In reality, the challenge was immense. No solids had been designed, let alone built, rated for human space flight and all the enhanced reliability and safety criteria that implied, and none had ever been designed for reuse.

colossus and neither the military nor NASA could effectively use such a behemoth. By this time the Saturn V, with 7.5 million pounds of thrust, was about to perform, and thrust-for-thrust the solid was just far too inefficient to equal the Saturn's lifting potential with its all-liquid rocket motors. No use, that is, until in 1971, NASA began looking at solids as a cheap alternative to liquid propellant boosters for the Shuttle.

But big solids had been developed for the US Air Force Titan III satellite launcher and that was a useful precedent for applying them to the Shuttle. Early in 1972 the decision was made to use solids over liquid boosters, but recover them by parachute and refurbish them so that they could be refilled with propellant and used again. And the reasons were all down to cost. Not only were they cheaper to build and fly, the cost of a lost liquid booster would be much higher than the loss of an equivalent solid. All the way down the calculations, solids provided the reduction in

The final offers from four contractors were lodged with NASA on 15 October 1973. Lockheed, Thiokol and UTC proposed segmented boosters comprising a series of barrel-shaped sections, each separately filled with propellant and taken to the Cape on rail cars where they would be bolted together in a vertical position in the Vehicle Assembly Building. Aerojet General had a lot of experience with testing solids and wanted a monolithic design – a non-segmented cylinder for safety and reliability fabricated as one complete structure. The big problem with that was the 500-ton weight of the monolithic booster, which would render it almost impossible to handle on anything other than massive barges across limited distances. Manoeuvring a 500-ton structure off the barge and into a vertical position would be difficult at best.

With their mighty monolithic concept, Aerojet virtually wrote themselves out of the race, although they did claim their design was the safest possible way to build SRBs for the Shuttle. A series of evaluation teams examined the remaining contenders, awarding technical points ranking Lockheed, Thiokol and UTC in order of preference. With very little between the top two, Thiokol's bid came in cheaper than Lockheed so they got the contract, announced on 20 November 1973.

In reality there were to be three contractors for the SRBs. Technically, the Thiokol contract was just for the Solid Rocket Motor (SRM), by far the greatest bulk of the structure since the motor was itself the booster. But on 21 December, 1973, United Space Boosters, a part of United Technologies, was awarded the contract for the Solid Rocket Booster, defined as the SRM from Thiokol plus the non-motor related equipment, as well as final assembly of all the elements. Then NASA's Marshall Space Flight Center (home of the Saturn launch vehicles and manager of the Skylab space station but now running low on work) was given the job of providing the recovery system, the separation thrusters and the electronic control units which, since it was 'in-house' work and not contracted out, also skimmed a little more off the cost.

SRB design and operation

Each SRB has a total length of 149ft 2in and a diameter of 12ft 2in and weighs about 1.25 million lb fully loaded with propellant. Each SRB used for the first seven missions had an average thrust of 2.8 million lb but, being solid propellant, a relatively low specific impulse of 268sec. From the eighth flight, launched on 30 August 1983, the thrust was increased to 3 million pounds as a result of development underway since 1980 and this is the type of motor that has been used since.

Each SRB is filled with propellant hardened to a solid, with a tunnel running down the centre of each booster. When ignited, the propellant will burn from the inside of the tunnel out towards the cylindrical wall, with all the energy going down the tube and shaped by the exhaust emerging from the bell nozzle at the bottom. The propellant will not burn through

the cylindrical wall of the casing because the latter is lined with a thrust inhibitor, which will terminate combustion.

The role of the SRBs is to provide 71 per cent of the lift-off thrust, and for the first 2min 3sec or so of flight. At the end of that time the Shuttle will have been propelled by the combined thrust of the SRBs and the SSMEs to a height of around 150,000ft. At that point the SRBs have expended their propellant and are separated from the External Tank, which continues to provide propellant for the three Space Shuttle Main Engines for an additional 6min. After depletion and separation, the SRBs continue upward on the momentum of their energy at cut-off, reaching a peak height at 220,000ft just 1min 15sec later, from where they begin to fall to the Atlantic Ocean below, approximately 141 miles down range of the launch site.

The SRB is made up from several separate sections including the nose cone, the Solid Rocket Motor and the nozzle assembly. The SRM is by far the biggest part of the booster, comprising eleven cylindrical D6AC steel sections. When assembled vertically, one on top of the other, they make a steel tube 116ft long weighing a total 150,023lb empty of propellant. The separate sections are each joined to the next by a tang and clevis joint, and secured with 177 pins spaced equally around the cylindrical section.

The eleven sections are stacked together making four separate cylindrical casting segments into which the propellant will be poured to make up the four motor segments,

RIGHT It is the SRBs that support the ET, to which the Orbiter is attached, and the entire stack stands on hold-down structures that keep the Shuttle anchored while the three main engines run up to speed prior to SRB ignition. The hold-down posts are severed by explosive bolts. *(NASA)*

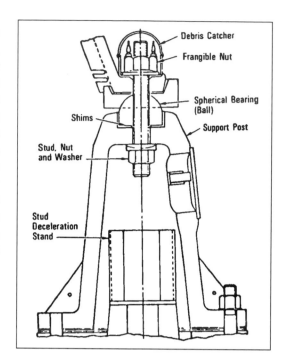

Debris Catcher

Frangible Nut

Spherical Bearing (Ball)

Shims

Support Post

Stud, Nut and Washer

Stud Deceleration Stand

known as the forward, forward centre, aft centre and aft motor segments. The upper three segments are each composed of two sections joined together, while the aft motor segment is assembled from three sections. These 'factory' joints are permanent, but mating the four separate segments into a full SRM will take place only in the Vehicle Assembly Building at the Kennedy Space Center, when the booster is assembled for launch at the 'field' joints where the four segments are joined.

The next step in preparation for delivery to the launch site will be to fill each of the four segments with propellant. Each SRB will carry 1.106 million lb of TP-H1148, which is a composite of 16 per cent aluminium powder as the fuel, 69.6 per cent of ammonium perchlorate as the oxidiser, 0.4 per cent of iron oxidiser powder as a burn-rate catalyst and 12 per cent of a rubber-based binding agent known as PBAN (polybutadiene acrylic acid acrylonitrile terpolymer) together with 1.96 per cent of an epoxy curing agent. The lining on the cylindrical steel walls of the SRM is UF-2137, an asbestos-filled carboxyl terminated polybutadiene polymer, which will protect the cylindrical walls.

Each of the four motor segments is filled with the propellant after first placing a shaped mandrel in the centre position of the cylinders. The propellant is poured around the mandrel

and will solidify so that when the mandrel is withdrawn there is a hole in the centre. When the segments are stacked together they will form a central tunnel running right down the middle. Booster segments are filled in pairs so that they come from the same mixing preparation. Any minor irregularities in thrust will therefore be experienced by both boosters, one each side of the Shuttle, to minimise asymmetric thrust which would have alarming consequences, mimicking the effect in an aircraft of having engines of unequal thrust under each wing, the more powerful engine tending to turn the aircraft round.

The central mandrel is not a cylinder, but is in the form of an 11-point star in the forward motor segments and a double truncated cone in each of the aft segments. When the propellant sets hard and the mandrel is withdrawn, the propellant will retain the mirror image of these different shapes. The thrust is controlled by the shape of the burning area and the different shaped tunnel areas control the thrust as the booster consumes its propellant. Because the weight of the Shuttle gets considerably lighter as the massive quantity of solid propellant is consumed, the boosters must progressively produce less thrust to limit maximum dynamic pressure to 650lb/sq and acceleration to no more than 3g. To do this, the propellant profile is shaped to reduce thrust by around 35 per cent at 55sec into the flight.

Because the SRBs are so heavy, any weight-saving measures are eagerly pursued and lightweight cases were developed with 0.003–0.005in removed from the cylinder walls, saving 5,000lb overall. First used on the sixth and seventh Shuttle flights in April and June 1984, there was concern that too much metal had been skimmed from the SRB sections. The Shuttle reverted to standard weight cases while an interim medium weight case was developed with walls 0.001–0.002in thinner, first used on the eleventh Shuttle flight in February 1984. Several flights used a secondary version of the lightweight case, skimmed 0.002–0.004in thinner and saving 4,000lb mass overall, and these were used throughout the remainder of the flight programme. These weight improvements helped immensely with clawing back excess mass that was difficult to eliminate from the Orbiter. Excess weight was always a problem with the Shuttle and never went away.

As related previously, the permanent joining of the separate sections into four casting segments are known as factory joints, leaving the field joints to bring the separate segments together, stacked and bolted at the Kennedy Space Center. The field joints consist of similar tang and clevis joints to the factory joints with two fluorocarbon O-rings to prevent the flow of gases from escaping through the join. This is possible because there is a slight gap between the propellant segments. The integrity of the joints is extremely important because any gases leaking between the segments will erode the gas-tight path and begin to melt the outside of the booster, or worse, impinge upon the External Tank and its volatile propellants. This is what happened to *Challenger* on 28 January 1986, when hot gas blew past the O-ring in the tang and clevis joint and opened a path like a blow-torch. The resulting flame burned a hole in the side of the External Tank, igniting the liquid hydrogen inside and causing an explosion which blew apart the boosters from the ET and broke the Shuttle into several separate sections, killing the crew on impact with the water.

It need not have happened. Boosters from several earlier missions indicated that the design was weak and that black smoke had been observed leaking from field joints, a vulnerability made much worse because the air temperature at launch of *Columbia* was below freezing. The rubberized O-ring was stiff and resistant to the pliability it needed for fast reaction to the pressure wave that should have sealed it. As a result of that catastrophe, the design of the tang and clevis joints was improved and a third O-ring added. Many small engineering changes were also made and when the Shuttle resumed flight operations it was with a much safer SRB design. Moreover, a weather seal has been added to prevent moist air intruding and heaters are added to keep the temperature at no less than 75°F.

The cylindrical forward skirt of the SRB sits on top of the forward motor segment and above that is the frustrum, a truncated cone incorporating the forward External Tank attachment ring and the main parachutes. It also contains four of eight separation thrusters carried by each booster – solid rocket motors that help to shunt the SRB aside from the ascending stack of External Tank and Orbiter. The conical nose section sits on top of the frustrum and houses the pilot and

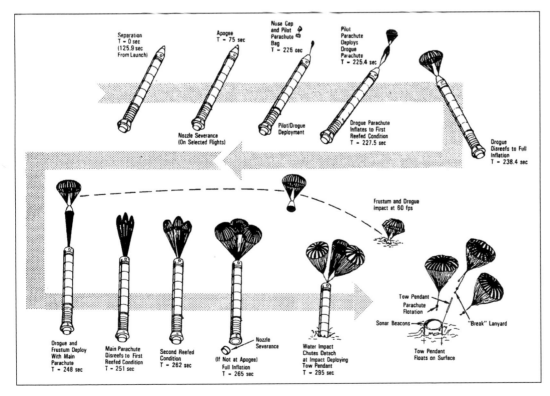

LEFT **Recovery of the SRBs allows them to be cleaned out and used again. Segments can be mixed in recycling, but when filled with solid propellant only those filled from the same batch will fly together.** (NASA)

drogue parachutes released prior to the main parachutes.

At the other end of the booster, the aft ET attachment ring completely encircles the SRB and is attached to the aft motor segment. Surrounding the exhaust nozzle of the SRB is the aft skirt, which ties down the entire Shuttle stack to the launch pad until it is released for flight. It also supports the other four solid thrusters to shunt the aft end of the SRB away from the Shuttle after separation. The nozzle is gimbaled like the Orbiter's liquid propellant SSMEs and is cooled by a carbon cloth that ablates continuously during firing. Each nozzle can move 8° in each of two axes, controlled by hydraulic servo-actuators.

The SRB gets its electrical power from the

Orbiter main dc bus power provided by the fuel cells with a nominal power level of 28 volts. Each SRB also has two completely autonomous hydraulic power units (HPUs) comprising an auxiliary power unit and associated fuel and pump systems, with the APUs fuelled by hydrazine to produce the shaft power to drive the hydraulic pumps. These are situated on the aft skirt with 22lb of hydrazine stored in each fuel tank pressurised with gaseous nitrogen to 400lb/sq in. Each HPU is linked to both servo-actuators and the APU has a maximum shaft speed of 80,640rpm, delivering a pump speed of 3,600rpm with pressure up to 3,750lb/sq in.

Ignition of the SRB is effected by an igniter 45in long containing a solid propellant that when fired sends a flash down the central tunnel to ignite the main propellant. The command is given when the three Orbiter SSMEs have reached 90 per cent thrust and no fail commands have been sensed. Three signals must be sent simultaneously for the igniter to fire, all commands coming from the SSME main engine controller. At the same time, the four hold-down bolts securing each SRB to the pad are severed. Each bolt is 28in long and 3½in in diameter. When the boosters ignite and the bolts are severed, the Shuttle will lift free from the launch pad.

Bringing it back

The burn duration of the SRB is set by the grain and the density of the solid propellant, while the thrust level is adjusted by the interior shape of the burning surface. At the end of the burn phase the SRB has exhausted all its propellant and the thrust falls off almost immediately. Pressure sensors activate the separation sequence when they detect a pressure drop below 50lb/sq in, at which point only trace gases are venting out of the exhaust nozzle. As the forward ball and socket and three aft struts are pyrotechnically blown, the eight thrusters – four forward, four aft – fire for 1.02sec to shunt the boosters away from the External Tank. The dramatic effect of the bright flame still pouring from the exhaust nozzle at the base of the booster and the sudden flash of fire fore and aft at separation creates an unexpected effect, but this is the standard

display and reassuringly normal. An interesting addition to the Orbiter's computer software has the Orbiter's nose thrusters fire momentarily to deflect the exhaust from the forward SRB separation thrusters and prevent it staining the outside of the crew compartment side windows.

As the SRB tops out its trajectory at 220,000ft and falls toward the Atlantic Ocean, a high-altitude barometric switch triggers ejection of the nose cap so that the 11ft 6in conical ribbon pilot parachute can deploy. This pops out at 15,700ft about two minutes after separation from the External Tank. The pilot 'chute pulls out the drogue pack from the frustrum approximately 30sec later, the 54ft drogue 'chute deploying on twelve 95ft suspension lines. Reefed initially, its job is to stabilise the booster in a vertical position and at 5,975ft, about 4min 8sec after separating from the ET, the frustrum is separated from the forward skirt and the suspension lines on the main parachutes are deployed. The lines are 204ft long and deploy the main 'chute to its initial, reefed condition at 4min 11sec, followed by a second reef 11sec later to minimise the shock of deceleration. Full inflation of all three 136ft diameter parachutes takes place at 4min 45sec. By this time the main 'chutes have slowed the booster from 230mph to 50mph, at which speed they impact the water approximately 4min 55sec after separation.

When the SRB enters the water it only partially fills with water, and when the parachutes are released the booster has a natural flotation characteristic that keeps it floating with the forward end 30ft above the water and the nozzle section below. A radio transponder communicates the location of the booster to a distance of 10 miles and a flashing light with a night-time range of almost 6 miles. The boosters are recovered by two special vessels, the *Freedom Star* and the *Liberty Star*, each 176ft long and with a 37ft beam, with a displacement of 1,052 tons. At splashdown the retrieval ships are on station approximately 10 miles from the recovery area, one booster to be recovered by each ship. Divers help recover the parachutes and the 5,000lb frustrum before turning to the booster shells.

Using inflatable boats, eight divers deploy to each booster and install a diver-operated plug,

(or DOP) which is a plug device with a length of 22lb and a weight of 1,100lb. The DOP has neutral buoyancy and is therefore relatively easy to manoeuvre in water. The divers manipulate the DOP down to the bottom of the booster, lying at an angle in the water, and insert it into the exit nozzle of the SRM approximately 110ft below the surface. Using powerful pumps on the retrieval vessel, air replaces water inside the long tube-like SRM and it begins to rise higher above the water, finally tipping over and floating horizontally on the surface. Attached by two lines, the two boosters are brought to Port Canaveral and pulled alongside the retrieval vessel where they are ready to be brought ashore. In due course the boosters are separated into their various segments, cleaned,

BELOW The lower skirt assembly separates during the descent on a separate parachute, leaving the nozzle looking decidedly wasted! *(NASA)*

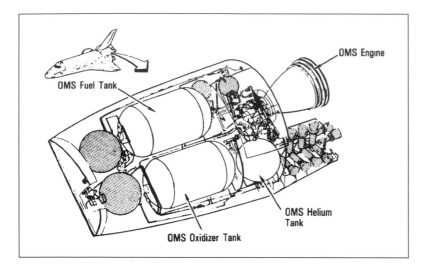

OMS Engine

OMS Fuel Tank

OMS Helium Tank

OMS Oxidizer Tank

composite materials. For various reasons, not least because of resistance at NASA to anything that would slow the pace at which Shuttle missions were being flown, these schemes fell prey to budget trimmers and bureaucrats – as the author knows during one sparring session at NASA headquarters! There is, however, one set of filament wound booster sections at the Space and Rocket Center, Huntsville, Alabama.

Orbital Manoeuvring System (OMS)

refilled with propellant and reused on another flight. Segments can be found intermixed with segments from another flight and may never come together again for a later launch.

The whole concept of booster design, development and partial reuse is a very long way from the original idea of a booster built like a giant airliner, carrying a reusable Orbiter and flying back to a landing for immediate return to the air for another mission. At several times since the first Shuttle mission, efforts have been made to introduce a much more reliable and lighter booster using filament-wound

Once it is inserted into orbit, the Shuttle has no need of any rocket motor or other form of propulsion to keep going round and round the earth. But the Orbiter's three main engines cut off prior to achieving the correct speed necessary to stay in orbit, at which point it sheds the External Tank and fires the Orbital Manoeuvring System (OMS) rocket motors to make up the extra speed to achieve orbit. The two OMS motors are contained in separate pods located on the upper aft fuselage either side of the vertical tail. They will provide all the thrust necessary for getting into orbit, making

orbital changes as may be necessary for activities such as rendezvous and docking, and for the vital means of returning out of orbit to a controlled descent through the atmosphere. The pods also contain the aft Reaction Control System (RCS) clusters for attitude control.

Each pod has a length of 21ft 10in and a width of 11ft 5in at its forward end, and 8ft 5in at its aft end. Eleven separate bolts lock each pod to the 'shoulder' position on the aft fuselage of the Orbiter. Each pod carries a load-bearing thrust structure fabricated from 2124 aluminium with cross braces in aluminium tubing and attachment fittings in the same metal. The forward and aft support bulkheads and floor truss beam are similarly machined from aluminium 2124. The centre line beam is in 2024 aluminium sheet and carries titanium stiffeners and graphite-epoxy frames. Skin panels of graphite-honeycomb sandwich construction cut the weight by 450lb over the two pods. Exposed pod surface areas are covered by the same thermal protection materials used on other areas of the Orbiter.

The functional purpose of each pod is to provide a support for the OMS engine and the tanks and propellant lines necessary for its operation. The two propellants are in separate

ABOVE After servicing is completed, the starboard OMS pod is lowered into position, where it will be attached to the right shoulder of the aft-fuselage structure. Note the foil thermal blankets. *(NASA)*

LEFT Having been serviced and cleared for flight, the two OMS pods are moved into their respective positions flanking the Orbiter's vertical tail, before the the three main engines are fitted. *(NASA)*

tanks within each pod together with a gaseous helium pressurant tank for pushing the fuel and the oxidiser to the OMS motors. Each engine produces a thrust of 6,000lb at a specific impulse of 313sec with a chamber pressure of 125 lb/sq in and a nozzle expansion ratio of 55:1. Design philosophy focused on simplicity and reliability and for that reason the propellants are hypergolic, which means they ignite on contact and dispense with the need for ignition systems. The OMS engines are vital for getting into orbit, conducting the mission and for getting back to earth so they must work on time, every time they are needed.

The propellants comprise 7,773lb of nitrogen tetroxide as the oxidiser and 4,718lb of monomethyl hydrazine (MMH) as fuel, brought together in a mixture ratio of 1.65:1. Built by Aerojet, each engine weighs 305lb and together they can apply a velocity change of 2ft/sec/sec, which, with the amount of propellant carried, a total installed velocity change of 1,000ft/sec. About half of this is used to push the Orbiter into orbit following separation of the External Tank at suborbital velocity. For most manoeuvres the OMS motors are used one at a time to reduce the number of starts and,

consequently, increase the life expectancy of each engine.

Reaction Control System (RCS)

While the OMS engines are used to make major velocity changes, attitude control for roll, pitch and yaw are made using a collection of thrusters in three modules: one in the upper forward fuselage and one in the rear of each OMS pod on opposite shoulders of the aft fuselage. The Reaction Control System (RCS) is made up from 38 prime thrusters and six secondary thrusters manufactured by the Marquardt Corporation, a company that had provided all the manoeuvring thrusters for the Apollo spacecraft and the Lunar Module. Like the OMS motors, the RCS thrusters used the same hypergolic propellants (N_2O_4/MMH) but with dedicated propellant tanks.

The Forward RCS (FRCS) module is fabricated from 2024 aluminium alloy skin-stringer panels and frames, the latter riveted to the skin-stringer panels. The FRCS is attached to the forward nose section by 16 fasteners. It

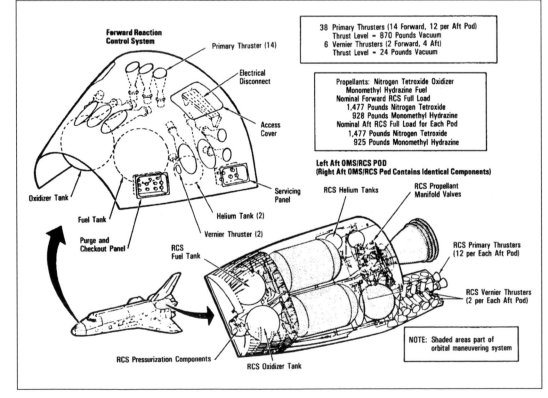

RIGHT All attitude control manoeuvres in space are made using thrusters in separate clusters located in the upper nose of the forward fuselage, and in cantilevered arm structures on the aft end of each OMS pod.
(NASA)

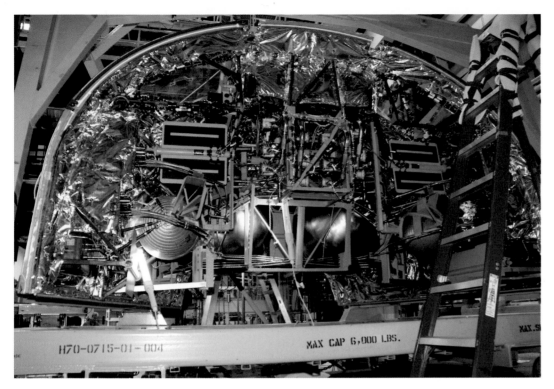

is a separate module for purposes of servicing and maintenance duties and contains 14 primary and 2 secondary thrusters. The RCS thrusters situated on each of the rear OMS pods are clustered along a cantilevered section aft of the main structure. Each aft pod arm carries 12 primary and two secondary thrusters. Each of the three RCS modules has a 1,477lb capacity oxidiser tank, a 928lb capacity fuel tank and two helium tanks for pressurising the propellants. The propellant tanks are each 39in in diameter with dry weights of 70.4lb for the forward tanks and 77lb for the aft tanks. Each helium tank is 18.7in in diameter with a volume of 3,043cu in and a dry weight of 24lb.

The 38 primary R-40A thrusters have a thrust of 870lb with a specific impulse of 289sec, capable of being operated for a minimum of 1sec or a maximum of 150sec. The design specification of each engine is for a total of 50,000 starts and 20,000sec of cumulative firing. The six secondary R-1E-3 motors each produce a thrust of 24lb at a specific impulse of 228sec. They can be operated between 1sec and 125sec, designed for a maximum of 330,000 starts over a total cumulative firing duration of 125,000sec. While the principal function of the RCS is to provide attitude control, the thrusters can be used

for tiny velocity changes, and for manoeuvres where changes of less than 6ft/sec (4mph) are required the RCS thrusters are used. In case of emergencies there is an interconnect system between the RCS and the OMS.

BELOW Seen here being carefully lowered into place, like the OMS pods, the forward RCS module will be filled with propellant at the pad. *(NASA)*

Chapter Five

Flying the Shuttle

Launched like a rocket and brought back to earth like an aircraft, this 100-ton glider routinely slices through the outer atmosphere at twenty-five times the speed of sound. Trimming more than 17,000mph from its speed in orbit, it floats down gently at less than 300mph to a runway 5 miles long, 5,000 miles distant from the point at which it began its unpowered freefall to earth. Like no other piloted aircraft ever built, the Shuttle is unique – one of a kind – and to fly on it is to reach the pinnacle of the flying experience. In 135 missions over more than 30 years, five Orbiters have carried more than 340 people into space.

LEFT Launched on 15 July 2009, *Endeavour* flies STS-127 to the International Space Station on the thrust of 3,500 tons – equivalent to the power of about 150 jet combat aircraft, or roughly 18,000 Battle of Britain Spitfires! *(NASA)*

ABOVE The Kennedy Space Center includes the two launch pads (A at top right and B at left) more than three miles from the giant Vehicle Assembly Building (VAB), at centre. At an angle to the right of the VAB is the Launch Control Center, or LCC. *(NASA)*

RIGHT The VAB has two exit/entry doors on each of two opposing sides, but Shuttle bays use the two nearest the launch pad. Just visible to the left is the Orbiter Processing Facility (OPF). *(NASA)*

Getting ready

Getting the Shuttle ready for a flight begins with the recovery of the two Solid Rocket Boosters immediately after the previous launch. Towed back to Port Canaveral by one of two recovery vessels, the empty solid rocket motor is separated into its major segments which are then inspected, cleaned and shipped to the manufacturer's plant in Utah for refilling with propellant. External Tanks will arrive by barge and be stored in the Vehicle Assembly Building (VAB).

On returning from space the Orbiter will land on a special runway at the Kennedy Space Center. It will then be taken to the Orbiter Processing Facility (OPF), a building consisting of two Hugh Bay areas, each 197ft long by 150ft wide and 95ft high, connected by a Low Bay, which is 233ft long, 95ft wide and 25ft high. It is here that Orbiters are turned around for the next mission and where inspection and replacement of thermal protection tiles is only one among several hundred different systems and subsystems that are examined.

If an Orbiter is to be withdrawn from the line for major maintenance it will go to the Orbiter Modification and Refurbishment Facility (OMRF), northwest of the VAB. Completed in 1987, it consists of a single High Bay area such as that in the OPF, together with a two-storey Low Bay. South of the VAB, a Logistics Facility provides a 324,640sq ft spares storage plant where some 190,000 separate items are kept for supporting work in the OPF and the OMRF. It has fully automated retrieval and handling equipment, allowing managers, engineers and technicians to keep track of every component and to maintain full history of its life during and after manufacture.

The VAB is the building where all the Shuttle elements converge and where the stack is assembled for rollout to the pad. It is one of the largest buildings in the world and has an interior volume of 129.5 million cu ft – big enough to contain all the pyramids of Egypt! The VAB is 525ft tall, 716ft long and 518ft wide, consisting of a High Bay area to the full height of the building and a Low Bay area 210ft high. There are four High Bay areas, but only two are

LEFT During the Apollo programme for which the facilities were built, the Launch Control Center housed 450 people involved in getting a Saturn launch vehicle off the pad. *(NASA)*

used for Shuttle stack integration. A legacy of the Apollo days, the VAB was built in the early 1960s and has been used for all Apollo and Shuttle assembly activity since the first Saturn V flight vehicle was rolled out in 1967.

When the SRB segments arrive back from Utah they are placed in the SRB Rotation and Surge Facility before being moved into the VAB, where they will be individually stacked and bolted together on a Mobile Launch Platform (MLP) in High Bay 1 or High Bay 3. High Bays 2 and 4 are used to store External Tanks, which will have been brought by barge from the manufacturer, Lockheed Martin (formerly Martin Marietta), in Louisiana. When the two SRBs are assembled, the relevant External Tank will be moved from its High Bay storage and lowered down between the two boosters and attached ready to receive the Orbiter.

The Orbiter is towed around from the OPF into the transfer aisle between High Bay areas where it is rotated 90° to a nose-up position using cranes within the VAB. There are more than 70 lifting devices in there from very small

RIGHT The ET is brought by barge from the Michoud Assembly Facility, Louisiana, down the Mississippi and round the tip of Florida to the Kennedy Space Center, where it is offloaded in the turning basin south-east of the VAB *(NASA)*

cranes to two 250-ton capacity giants. It used to be said that to qualify for a job here, a crane driver had to lower a crane hook and trap a fresh egg against the floor without cracking the shell. This is an urban myth – but all crane drivers can do just that, as they proudly boast!

Ever since 1967, the MLP is the transportable base upon which all Saturn moon rockets and all Shuttle vehicles have been stacked. It is a steel structure 160ft long, 135ft wide and with a height of 25ft. It weighs 8.23 million lb (4,000 tons) and will support the Shuttle stack as it is conveyed to the launch pad. When the External Tank is fuelled the complete MLP/Shuttle will weigh 12.7 million lb (6,300 tons). In the VAB and at the pad, the MLP and its load rest on just six 22ft-tall pedestals, through which that enormous weight is supported. Six pedestals are also provided outside the VAB to support the MLPs when not in use.

The MLP has three square-shaped openings for exhaust gases: two for the SRBs are each 42ft long and 20ft wide; the single cut-out for

ABOVE The Crawler Transporter moves the Mobile Launch Platform back from one of the Shuttle launch pads, just visible at left. *(NASA)*

LEFT The world's largest self-propelled vehicles, the Crawler Transporters were built in 1965 by the Marion Power Shovel Company at a cost of $14m each. One is called *Hans*, the other *Franz* after the two comic bodybuilders on TV show *Saturday Night Live*. *(NASA)*

133

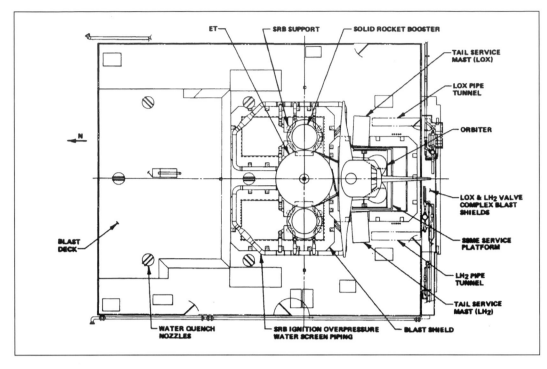

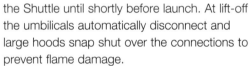

the three Orbiter SSMEs is 34ft long and 31ft wide. Two large structures called Tail Service Masts, each 15ft long, 9ft wide and 31ft tall, appear to encapsulate the inboard aft-fuselage, carrying umbilical feed lines for liquid hydrogen and liquid oxygen into the External Tank. Other service lines feed helium and nitrogen, while electrical connectors carry ground power to

the Shuttle until shortly before launch. At lift-off the umbilicals automatically disconnect and large hoods snap shut over the connections to prevent flame damage.

To facilitate assembly of the various Shuttle elements the MLP is moved into the VAB, and from there out to the launch pad on a massive Crawler Transporter (CT) that is 131ft long, 114ft wide and 20ft high, weighing 6 million lb (3,000 tons) unloaded. The CT moves on four double-tracked crawlers, each 10ft high and 41ft long, powered by two 2,750hp diesel engines driving four 1,000kW generators providing electrical power to the 16 traction motors which, operating through gears, turn the crawler tracks. Each of the 57 shoes on each track weighs 1,984lb. The CT is driven from one of two cab positions at diagonal corners of the vehicle. Its fuel tanks hold 5,000 US gal of diesel and the engines consume 126 gal/mile!

After the Shuttle stack has been assembled on the MLP and the CT positioned underneath, the entire assembly slowly moves out of the VAB and along to the pad, Launch Complex 39, more than three miles away. With the MLP and the empty Shuttle on its back, the moving stack weighs about 14.8 million lb (7,400 tons) – the world's heaviest tracked vehicle. The Crawlerway is as wide as an eight-lane motorway and consists of two 40ft-wide lanes

LEFT The Mobile Launch Platform supports the Shuttle on hold-down posts below each SRB, while umbilical and electrical connectors are encapsulated by the two tail service masts adjacent to the upper surface of the Orbiter's wings. *(NASA)*

FAR LEFT Two paired SRBs stand erect on a Mobile Launch Platform in a High Bay inside the VAB. *(NASA)*

LEFT With the height of a 16-storey building, the ET looks diminutive suspended in storage in the VAB, where it will be mated to the SRBs before receiving an Orbiter. *(NASA)*

RIGHT Threading the ET down through the multiple access platforms that will be used by technicians attaching it to the SRBs. (NASA)

FAR RIGHT *Endeavour* **in the Orbiter Processing Facility, where all Orbiters are serviced and sealed for moving to the VAB and stacking.** (NASA)

BELOW *Discovery* **shows signs of several previous missions, as tiles on the forward fuselage display streaking from hot plasma on the nose of the vehicle.** (NASA)

separated by a median strip 50ft wide. The CT moves at 2mph (1mph around corners) and has a levelling system for raising the top deck to lift the MLP off the support pedestals in the VAB, keeping the stack vertical as it ascends the shallow ramp on to the pad, where it is lowered to another set of pedestals ready for the countdown. The CT then withdraws and returns to the parking area near the VAB.

Launch Complex 39 consists of two separate launch pads, LC-39A being 3.44 miles from the VAB and 48ft above sea level, and LC-39B 4.24 miles away and 55ft above sea level. When the entire area was laid out by NASA as the launch site for its giant Saturn rockets, four pads were thought to be needed but that was reduced to three, with eventually only two being built. LC-39A was converted from Saturn to Shuttle operations in 1978 and was used from the first Shuttle flight in April 1981. Conversion of LC-39B was completed in 1985 and was used first by the ill-fated STS-51L mission of January 1986 in which *Challenger* was destroyed shortly after launch. Most Shuttle flights began from LC-39A, the last mission from LC-39B being the STS-116 mission in December 2006.

ABOVE Received in the VAB, *Atlantis* is rotated 90° prior to mating with the ET. Note the engine covers which will be removed at the launch pad prior to flight. *(NASA)*

LEFT Lowered into position for attaching to the ET, *Atlantis* will have been scrupulously checked for loose items, which could fall away during rotation or float free in weightlessness to lodge in critical components. *(NASA)*

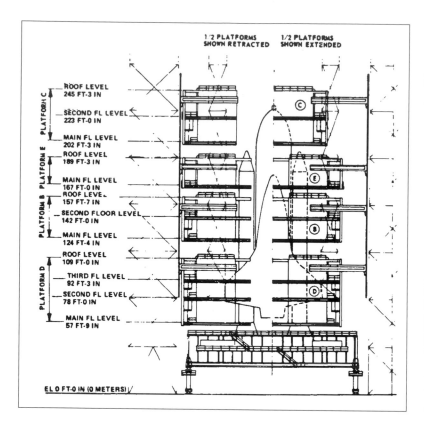

LEFT The design plan for the access platforms in the VAB reveals the extent to which technicians can reach almost all areas of the exterior. *(NASA)*

BELOW The Shuttle moves out to the pad. At left is the turning basin where ETs arrive from their assembly plant in Louisiana, and just to the right of the VAB are the Orbiter Processing Facility and the Orbiter Modification and Refurbishment Facility *(see page 132)*. *(NASA)*

ABOVE When suspect fuel pump tip seals were found on one of *Discovery*'s main engines from the preceding flight, the No.1 engine on *Atlantis*, scheduled for STS-101, was replaced after it had been stacked in the VAB – an unusual procedure working with the Orbiter in a vertical position. *(NASA)*

BELOW The 247ft-tall Fixed Service Structure west of the launch pad provides a secure mount for the Rotating Service Structure, which will swing through 120°, encapsulating the Shuttle to provide five work levels. *(NASA)*

ABOVE The sheer scale of the Crawler Transporter bearing the Mobile Launch Platform and its Shuttle load comes into focus at the pad. *(NASA)*

When the Shuttle arrives at the pad it will be prepared for launch and to achieve that a Fixed Service Structure (FSS) is permanently located alongside. It provides the main hinge for the Rotating Service Structure (RSS), designed to swing around and encapsulate the Orbiter for pre-launch work and to place payloads in the cargo bay. The FSS also carries an Orbiter Access Arm, at the end of which is a hermetically sealed chamber known as the 'white room' where the crew enter the Orbiter through the side hatch into the mid-deck area. This remains extended until about 7 minutes prior to launch, serving as an emergency escape route in the event of an impending fire or explosion.

The FSS also provides a rapid means of escape to the ground. It consists of a slide wire 1,200ft long, down which the crew can travel in specially designed baskets, each holding up to three people. At the bottom, some distance

LEFT Lockdown at the launch pad puts the Mobile Launch Platform down on six pedestals, which will support the full load until launch when it is returned to the VAB area. *(NASA)*

LEFT One of the corner bolts at the location of the interface between the Mobile Launch Platform and the Crawler Transporte on Launch Complex 39. *(NASA)*

RIGHT The levelling system incorporated into the Crawler Transporter maintains the load at vertical during the climb up to the launch pad. Note the 80ft-tall lightning mast atop the Fixed Service Structure. *(NASA)*

BELOW The bottom of the Rotating Service Structure (RSS) is 59ft above the concrete launch pad and extends to a height of 189ft above that level. Essentially a rotating bridge, the track traversing the flame trench dividing the pad provides a rail on which the RSS will move to enclose the Shuttle. *(NASA)*

ABOVE **The flame trench lies between the two sides of the crawlerway on top of the launch pad, bisected by the rail on which the RSS will swing across to the Shuttle.** *(NASA)*

away, an emergency shelter bunker provides a safe haven while the crew rides out any explosion that may take place on the pad itself.

The RSS is 102ft long, 50ft wide and 130ft high and rotates through 120°. It has five levels providing access to the Orbiter cargo bay and various platforms that can be extended to afford access to specific areas around the Shuttle. It also incorporates the Payload Change-out Room (PCR), which receives a Payload Canister with payloads loaded into what is basically a mirror-image of the interior of the Shuttle cargo bay. Consisting of a hermetically sealed box 65ft tall and 18ft 7in wide, it matches the interior dimensions and layout of the cargo bay. There are two Payload Canisters at the Kennedy Space Center and two special-purpose vehicles for moving them around.

Big payloads such as space station modules are kept in the Space Station Processing Facility (SSPF) and are placed inside the cargo bay when the Shuttle is in the OPF. But some equipment is installed via the Payload Canister with the Shuttle in a vertical position on the pad. Loaded up in the Vertical Cargo Processing Facility (VCPF), the Canister is trucked to LC-39

ABOVE With a width of 58ft and a length of 580ft, the flame trench is 40ft deep and will channel the exhaust from the three Orbiter main engines and the two SRBs in opposing directions. *(NASA)*

LEFT Down in the flame trench, the main flame deflector takes the form of an inverted 'V', 38ft high, 57.6ft wide and 40ft high. Faced with a volcanic ash aggregate, the outer coating of the deflector will be turned to glass by the heat of the flame. *(NASA)*

RIGHT Depending on their size, payloads are either placed in the Orbiter in the OPF or at the launch pad. Space station modules such as this Node are stored in the Space Station Processing Facility, but Spacelab modules were brought to the OPF for installation. *(NASA)*

RIGHT The Payload Canister (PC) brings payloads from the Vertical Processing Facility to the launch pad by truck. It mirrors the interior of the Orbiter payload bay, so the Canister can be hoisted up into the Payload Changeout Room (PCR), which will encapsulate the Orbiter when the RSS is rotated. *(NASA)*

FAR RIGHT With the PC installed in the PCR, the RSS can rotate round to mate with the Shuttle. *(NASA).*

where it is winched up into the vertical position, with the RSS in the open position. Once inside, the RSS closes around the Shuttle and is sealed around the edge as the payload bay doors are opened and the cargo transferred across from the Canister to the Orbiter. Leaving the payloads in the Orbiter, the process is then reversed as the RSS swings open and the Canister is removed.

Preparation for launch begins with securing all electrical and umbilical connections between the ground and the Shuttle, followed by loading of hypergolic propellants into the Reaction Control System (RCS) and the Orbital Manoeuvring System (OMS) tanks. Because these propellants ignite on contact with each other they are loaded in series and never in parallel. After this the cryogenic oxygen and hydrogen fluids for the Shuttle fuel cell electrical power production units are pumped into special tanks on the FSS. Then the vehicle is powered down so that the pyrotechnics can be safely installed – electrically activated squib devices

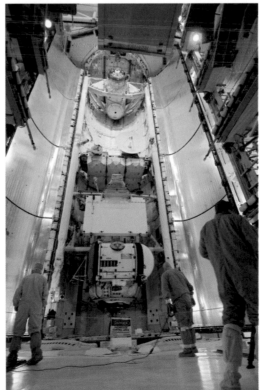

ABOVE After depositing its load at the launch pad, the Crawler Transporter drives back down the crawlerway to park back near the VAB. The PCR is empty. *(NASA)*

LEFT Inspecting a suite of Shuttle payloads destined for the International Space Station in the PCR area. The docking module is located at the top of the cargo bay. *(NASA)*

that will sever connections between elements at various stages of the ascent. It is now that crew equipment such as Extra-Vehicular Activity (EVA) space suits is installed in the Orbiter as well as personal items, which the crew has pre-selected, sometimes in discussion with their families.

The countdown is controlled from the Launch Control Center (LCC) alongside the Vehicle Assembly Building (VAB), a four-storey building 378ft long, 181ft wide and 77ft high. Built as part of the original Apollo-Saturn programme, the LCC has four firing rooms from where the countdown operations are maintained. In the Apollo days it took 450 people to get a Saturn V off the pad, but now just 90 are needed to launch the Shuttle. This is down to more efficient use of personnel and by the significant increase in automation and computerised processes. Moreover, a countdown takes around 40 hours, less than half the time taken to get the early Shuttle missions on their way in the 1980s.

Three days before launch the test conductor gathers his or her team and goes through the various station positions in the firing room. First, the software is loaded into the general purpose computers (GPC) aboard the Orbiter, and back-up systems in the mass memory units are checked and verified before being stored in the fifth GPC. Work platforms in the Orbiter are removed from the crew compartment and flight equipment installation completed. The Orbiter

fuel cell tanks are serviced prior to cryogenic loading and preparations begin for pumping propellant into the External Tank.

Two days before launch all non-essential personnel are cleared from the pad while liquid hydrogen and liquid oxygen are loaded to the fuel cell reactant tanks. When that task is completed the pad is opened up again to service personnel. They activate and check the flight control equipment, guidance and navigation systems and install seats in the mid-deck area for the requisite number of flight crew – usually seven. Switch configurations and control panels are checked and configured for receiving the flight crew, and preparations are made for rolling back the RSS. It is here that a built-in hold of up to 26 hours can be applied to check on specific payloads, special equipment unique to a specific mission or to catch up with uncompleted work.

One day before launch the countdown is resumed and the RSS is rolled back around the FSS, exposing the Shuttle to full view. The last few items of crew equipment are installed, the final checks are made of switch positions and samples of the air in the cabin verify a clean condition. The fuel cells are powered up and a purge helps verify their operating condition, providing electrical power to spacecraft systems. Then the pad is again cleared of non-essential personnel as gaseous nitrogen pours into the cargo bay and other cavities prior to loading the External Tank with propellant.

LEFT AND ABOVE Astronauts use a converted Grumman Gulfstream II to fly touch-and-goes up to and including launch day, to confirm weather conditions for winds aloft and visibility to check against abort criteria. For training purposes, the left side of the cockpit is configured like the Orbiter, the right is left in unmodified Gulfstream layout. *(NASA)*

BELOW LEFT Rehearsing weightless operations can be fun, but has a serious side. Astronauts cavort in the 'vomit comet', aircraft that plunges earthwards on a steep descent path to simulate weightlessness. *(NASA)*

BELOW RIGHT The traditional launch day breakfast, an informal gathering in the crew quarters before flight. *(NASA)*

Launch day

On launch day, filling the External Tank takes place while communication checks are made with various stations around the Cape Canaveral area and the gimbal movement of the OMS engines is verified. The inertial measurement unit (IMU) is calibrated and tracking antennas that will follow the Shuttle after lift-off are aligned.

T minus 5hr 20min – a built-in hold starts during which time an ice inspection team goes to the Shuttle to see that no ice is building up on the exterior of the External Tank due to the presence of the super-cold propellants.

In their quarters in the Operations and Checkout (O&C) Building the flight crew will eat a meal and get a weather briefing. Almost invariably, another astronaut will be flying touch-and-goes on the runway at the Kennedy Space Center to get a first-hand feel for the winds aloft and visibility, and this information will play a crucial part if meteorologists indicate marginal conditions.

T minus 2hr 30min – the flight crew leaves the O&C building and is driven to the pad where they access the Orbiter via the White Room. Establishing communications with the LCC Firing Room they also hook up to the Mission Control Center (MCC) at the Johnson Space

ABOVE **The flight crew depart for the pad and their last visual encounter with workers at the Kennedy Space Center, a final wave goodbye.** *(NASA)*

RIGHT Located at the 195ft level, the Orbiter Access Arm (OAA) is attached to the Fixed Service Structure, the RSS by this time having been retracted across the flame trench bridge. *(NASA)*

ABOVE Astronauts are conveyed to the launch pad in the Crew Transfer Van, a journey taken to Launch Complex 39 ever since Borman, Lovell and Anders left for their Christmas flight to the moon in December 1968. By the end of Shuttle flights, 153 rockets had been launched from these two pads. *(NASA)*

Center in Houston, Texas. Mission Control will monitor the countdown and launch and take over control of the mission from the point where the Shuttle ascends above the FSS.

After the access hatch is closed a cabin leak check is performed, the IMU alignment is made and the White Room is evacuated. As the close-out crew departs the pad to a fallback area, the primary ascent guidance data is transferred to the back-up computer.

T minus 20min – a 10-minute hold begins where last minute tasks are tidied up. When the count resumes the computers are switched to launch configuration and the thermal conditioning of the fuel cells begins.

T minus 9min – a second 10-minute hold starts and prior to resuming the count the Test Conductor polls the Firing Room for a 'go' condition on all screens. This is when the Ground Launch Sequencer [GLS] is switched on and the

LEFT The White Room is a box-like structure at the end of the OAA, and usually holds six people. It is here that the crew enters the Shuttle through the ingress/egress hatch on the port side of the forward fuselage, directly into the mid-deck area. *(NASA)*

RIGHT The commander occupies the left seat on the flight deck, the pilot to his right. Two mission specialists occupy lightweight seats behind, and to the right rear, but they still have tasks to carry out in the event of a malfunction. *(NASA)*

terminal countdown begins, where all functions from this point are automatic and controlled by the GLS computer in the Firing Room.

T minus 7min 30sec – the orbiter access arm is retracted, but should it be needed for an emergency evacuation of the Orbiter it can be back in place within 15sec.

T minus 5min 15sec – Houston sends a command to start the operational instrumentation recorders which will contain information on the vehicle during ascent, on orbit and during re-entry and landing. In essence, these are the Shuttle's black-box data recorders, data from which will be analysed in great detail after the mission.

T minus 5min – the Auxiliary Power Units (APU) are fired up to power the hydraulic system and the SRB firing circuits are enabled, as well as range safety systems that will detonate the boosters if they run amok and threaten inhabited areas.

T minus 4min 55sec – the liquid oxygen vent on the External Tank is closed. Until this time it has been boiling off oxygen turned to a gas to prevent excess pressure building up inside the LOX tank.

T minus 4min – a helium purge is conducted of the three Orbiter main engines.

T minus 3min 55 sec – the hydraulic systems move the aero-surfaces (elevons, speed brakes and rudder) to show working order.

T minus 3min 30sec – the fuel cells take over responsibility for all Shuttle electrical power and the ground system is disconnected.

T minus 3min 25sec – later by movement of the Orbiter main engine gimbal actuators to demonstrate effective working condition.

T minus 2min 55sec – the LOX tank is brought up to full flight pressure.

T minus 2min 50sec – the External Tank vent hood – known as the 'beanie-cap' – is retracted. Until this point it has been used to remove gaseous oxygen from the vent line and prevent ice building up around the orifice.

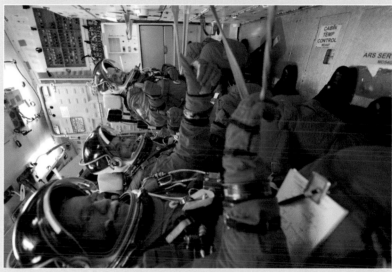

ABOVE Where more than four crewmembers are carried – which is most flights – the remaining astronauts are carried on lightweight seats in the mid-deck area. *(NASA)*

LEFT When the Shuttle is ready for launch, the RSS is rotated back and the pad is cleared. Since the loss of *Challenger* in January 1986, largely due to low temperatures, an ice team inspects the vehicle in the closing hour of the countdown to check for ice build-up. *(NASA)*

T minus 2min 35sec – the gaseous oxygen and hydrogen that had been supplied to the fuel cells from a ground source is disconnected and the flow is switched to the on-board reactants.

T minus 1min 57sec – the External Tank's liquid hydrogen boil-off vent is closed and the tank is brought up to flight pressure, but instead of being dispersed in the atmosphere as with the vented oxygen, it is piped away from the pad where it is burned off.

T minus 31sec – the terminal launch sequence starts. If any holds are called due to the slightest, minor, problem the countdown will have to be recycled to T minus 20min. Only one manual input is needed from this point, a 'go for main engine start' decision at about T-10sec. Up to this point, although the countdown has been proceeding on a pre-programmed computerised sequence, where several hundred functions are being monitored by the GLS, there is a considerable amount of human processing going on all around the world. Tracking stations, emergency landing strips, potential abort landing locations and weather in the abort recovery zones all must converge to show a favourable situation – and that is best done by human interaction.

T minus 28sec – the SRB's hydraulic power units are activated. These will be used for gimbaling the SRB nozzles for directional control, along with the Orbiter's main engines, during ascent.

T minus 16sec – the hydraulic power units move to prove workable authority. At this point, too, the sound suppression water system is switched on whereby 7,300gal/sec of water is poured across the exhaust vents in the Mobile Launch Platform to dampen the sound waves

from the Solid Rocket Boosters, preventing them from striking the Orbiter and damaging its wings and tail. Some damage was observed during the first launch on 12 April 1981, after which the sound suppression system was installed.

T minus 10sec – the 'go for main engine start' command is issued by the GLS, by which time the astronauts have received the good wishes of the launch team, lowered their visors, cinched the safety straps as tight as they can and braced themselves for one of the most violent rides of their lives.

T minus 8sec – the precise launch coordinates are transformed to a position vector. For LC-39A this is 28deg 36min 29.7014sec north latitude, by 80deg 36min 15.4166sec west longitude, the Orbiter's tail pointing due south. Now the GLS is analysing data points across the stack, watching for any off-nominal reading, monitoring the Shuttle systems as the vehicle literally breathes and groans as though it were alive. Sparklers are ignited beneath the three Orbiter main engines to burn off excess hydrogen under the exhaust nozzles and prevent a rough start and 0.5sec later the computers command open the valves to start the flow of liquid hydrogen and liquid oxygen into the turbo pumps on each SSME.

T minus 6.6sec – main engines 3, 2 and 1 ignite in that order at 120-millisecond intervals. Within three seconds they have ramped up to 90 per cent thrust

T minus 3sec – if the GPCs in the Orbiter give the 'go for launch' order, the SRB ignition sequence starts from which there is no bale-out. Come what may, the Shuttle is going to leave the pad. In the interval between ignition of the SSMEs and ignition of the two SRBs, the whole stack bends forward, pushed over 25½in at the nose by the shock of the engines firing. The gap between the two ignition points allows the Shuttle to return to vertical before it lifts off, preventing excess stress on the vehicle being pushed solidly upward before it has returned to a fully upright position. Early flights failed to anticipate this and the ignition sequence was backed up to accommodate the so-called 'twang'.

Lift-off

At ignition of the SRBs the Shuttle leaps off the pad with a noticeably higher rate of acceleration than the Saturn V, which produced only a little more thrust than the Shuttle but had a launch mass more than one-third greater.

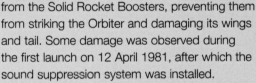

ABOVE LEFT Ignition of the two SRBs follows a systematic electronic checkout of Orbiter main engine status before the point of no return – when the solids fire there is nothing to stop the Shuttle leaving the pad. *(NASA)*

ABOVE At lift-off, both SRBs and the three main engines are at maximum thrust, but just 40 seconds after lift-off the SRBs will reduce thrust and the main liquid propellant engines will be throttled back to limit stress on the vehicle at Max-q. *(NASA)*

RIGHT Producing around 7.4 million pounds thrust at lift-off, with a thrust/weight ratio of about 1.7:1 the Shuttle fairly sprints off the launch pad compared to conventional expendable rockets – much faster than the old Saturn V launch vehicles with a thrust/weight ratio of 1.25:1. *(NASA)*

BELOW For the first 20 seconds of flight the thrust of each SRB is at a maximum 3 million pounds, then is shaped down by the solid propellant to 2.2 million pounds at 50 seconds (Max-q), increasing again to 2.5 million pounds by 75 seconds, before tailing off again down to 1.6 million pounds at 110 seconds, and a sharp decline to zero around 125 seconds. *(NASA)*

T plus 7sec – the tower is cleared, Houston takes over and the roll programme is initiated with the Shuttle at a speed of 127ft/sec (87mph). This turns the Shuttle through 180°, placing it in a heads-down attitude, aligning it with the desired orbital plane.

T plus 20sec – the Shuttle is in its correct roll orientation and at about 78° pitch, the Orbiter main engines tracking the inside of an imaginary curve toward orbit, the flight crew able to view the horizon upside down out the top of their forward windows. This attitude helps reduce wing loading through max-q (maximum dynamic pressure).

Rotation around its vertical axis places the Shuttle on the appropriate heading for its assigned orbit. On the first 25 missions, before the *Challenger* disaster of January 1986, the Shuttle launched commercial satellites which would be placed in a due east orbit of 28.5° inclination, but when the assembly of the International Space Station began in 1998, Shuttle flights would head for a 52° orbit so that the station's orbital ground track could cover all areas of the planet between those two lines of latitude, north and south of the equator. Because the Shuttle is prohibited from flying over populated areas of the United States while it is ascending to orbit, no flights can exceed a launch azimuth of 57°, since that would take it across the eastern seaboard.

T plus 26sec – the main engines are throttled back to about 55–60 per cent of rated thrust, and then back up again as the weight comes off due to combustion of solid and liquid propellants. The effects of ignition and launch, as sensed by the crew, is muted when the three main engines ignite, a solid rumble coming up through the structure more as distant thunder. But the sudden pitch forward as the 'twang' effect is felt is alarming if not anticipated. When the two boosters ignite there is no mistaking the effect and the smooth lift from the pad is quickly replaced by a further combination of potentially alarming sensations as the vehicle rolls on to its heading. Because the crew are placed off-centre from the centre of rotation, the effect is one of being wrenched around in a whirligig while simultaneously accelerating upward at an ever increasing pace only to be falling over backwards with the Orbiter hanging upside down in relation to the ground.

The two solid boosters give a rough ride at best and a bone-jarring experience at worst. The grain effect of the solid propellant mix gives a rough ride, where eyes find it hard to focus and the general sensation is of riding at relatively high speed over a road of broken cobbles in a vehicle without any form of suspension. Added to which, the flexing and twisting of the Shuttle's motion, as the engines gimbal to keep the vehicle on course in bumpy air streams and high wind, is a combination most astronauts find unique and quite difficult to forget!

If things go wrong

When the two boosters stop thrusting and separate at little more than two minutes into flight, the ride smoothes out but the danger of a potential abort remains with the crew all the way into orbit. There are essentially four abort modes, three of which are feasible and a fourth which is highly problematic. They all accommodate situations in which one or two main engines fail but with a varying degree of energy available.

The Transoceanic Abort Landing (TAL) is for situations where there is insufficient thrust to make it into orbit, but enough to push the Orbiter across the Atlantic Ocean so that it can drop the External Tank and descend to a landing at an assigned airfield in Western Europe or Western Africa – either Banjul in The Gambia, Ben Guerir in Morocco, or Zaragosa in Spain. This TAL capability usually extends from T plus 2min 50sec to T plus 6min 15sec for 28.5° orbits, or from T plus 2min 10sec to T plus 7min 45sec for high inclination orbits. It was originally known as the Transatlantic Abort Landing until the US Air Force protested that aborts of this kind on its military missions from the Vandenberg Air Force Base in California would take place over the Pacific and not the Atlantic. Since it cancelled plans to use the Shuttle before any flights from Vandenberg, the name change is largely academic.

The Abort-Once-Around (AOA) mode is used where there is insufficient energy remaining with either the three main engines, or the two orbital manoeuvring systems (OMS) engines to make it into orbit, so the entire trajectory becomes a depressed ballistic path where the Shuttle executes all the elements of a normal flight to orbit but starts descending on a normal re-entry before orbit is achieved. The Abort-To-Orbit (ATO) mode is more complex and gets the Shuttle into orbit, but lower than preferred, and this option begins at around T plus 4min 35sec for 28.5° orbits, or T plus 4min 7sec for 57° orbital inclinations. This method was used during the nineteenth Shuttle mission when one engine on *Challenger* shut down early after launch on 29 July 1985. Had a second engine failed the crew would have conducted a TAL abort.

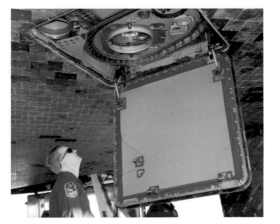

The fourth abort mode goes by the innocuous name of Return-To-Launch-Site (RTLS) and covers the loss of a main engine between SRB separation, around T plus 2min, up to approximately 4min 8sec when the Orbiter is at a speed of 8,200ft/sec (5,590mph). If an engine is lost in that period, no other abort mode is plausible. It involves the Orbiter continuing to burn propellants remaining in the External Tank while pitching down so that, flying upside down, it executes a loop that has the Shuttle, now in a heads-up attitude, pointing back toward the Kennedy Space Center. At propellant depletion the External Tank would separate and the Orbiter will make an approach as though returning from space, getting back down onto the ground approximately 25min after launch.

There are various options within the RTLS abort mode. If an engine should fail immediately after lift-off, the decision to go to RTLA would have that procedure begin immediately after SRB separation (around T plus 2min), but if

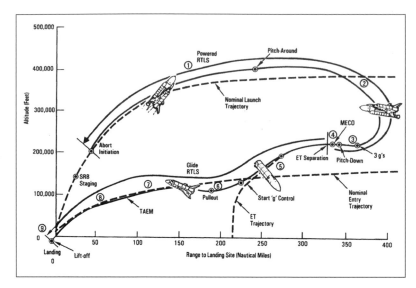

ABOVE A Return to Launch Site (RTLS) abort profile no astronaut would willingly choose to fly, with only a marginal chance of making it back to the runway at the Kennedy Space Center. *(NASA)*

BELOW Two SRBs float upright before divers secure them and pump out water to bring them horizontal for a tow back to Port Canaveral.

an engine should fail later the RTLS could be adopted out to the point where the other modes become possible. Whether the Shuttle could survive such a manoeuvre is uncertain and studies show there would be little margin for error. Moreover, if the Orbiter had to ditch there is no chance of it surviving intact with even a modest amount of cargo in the payload bay. Only by returning empty from a mission that left payloads in space would there be even a remote chance of the Orbiter surviving a landing on water, and even then only in very calm waters. Which is why the changes introduced after the loss of *Challenger* incorporated a bale-out option to abandon the Orbiter during descent if it was unable to make it to a landing strip *(see Chapter 3)*.

Into orbit

With the Shuttle progressing to orbit on a standard mission, the Orbiter would remain in a heads-down attitude so that the S-band communication system could maintain contact with the Bermuda tracking station. But since that facility has been retired the procedure has changed. Starting from late 1997, most flights involve the Orbiter rolling to a heads-up attitude after the Shuttle is outside the earth's atmosphere and wing-loading considerations no longer apply. The roll manoeuvre is performed at about T plus 6min at a speed of about 12,200ft/sec (8,320mph) and takes approximately 36sec to complete. This places the Orbiter in optimum attitude to achieve communications through the Tracking and Data Relay Satellite in geostationary orbit 22,300 miles above the earth.

All the way up the three main engines throttle back to keep acceleration below 3g until, at about T plus 8min 30sec, the three main engines shut down and the Shuttle is almost in orbit. The now depleted External Tank is separated and the forward and aft RCS thrusters manoeuvre the Orbiter away from the giant, cylindrical tank. Now the crew experience weightlessness and on the flight deck the pilots prepare to nudge the vehicle into a stable orbit.

Early in the Shuttle programme two manoeuvres were needed by the OMS engines to do that. The first, just a few minutes after main engine cut-off, known as OMS-1, would push the high point of the trajectory to a safe altitude where, half an orbit later, OMS-2 would circularise the orbit. Later missions have opted for a single burn into orbit from what is known as a direct-ascent trajectory, where the Orbiter's three main engines would steer a profile that incorporated the first OMS burn into the trajectory.

Checking for exterior damage

On reaching orbit a wide range of duties fall to every crew member. And for all, one of the most important is to perform a scan of the exterior surface of the Orbiter to determine any levels of damage to the thermal insulation so

ABOVE *Columbia* was destroyed by a piece of foam from the ET striking the wing leading edge and breaking an RCC panel. Since then, NASA has flown an Orbiter Boom Sensor System (OBSS) to inspect the exterior surface. *(NASA)*

ABOVE The OBSS comprises a 50ft long boom carried on the Orbiter along the starboard side of the payload bay on the opposite side to the remote manipulator. *(NASA)*

RIGHT The OBSS is picked up from the side of the Orbiter by the remote manipulator arm and manoeuvred around the exterior of the Orbiter using an Intensified TV Camera and a laser dynamic range imager. *(NASA)*

BELOW To be left aboard the International Space Station by the last Shuttle flight, the OBSS also has a hand rail so spacewalking astronauts can use tile repair kits on damaged areas of thermal protection material. *(NASA)*

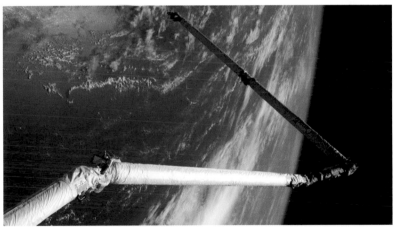

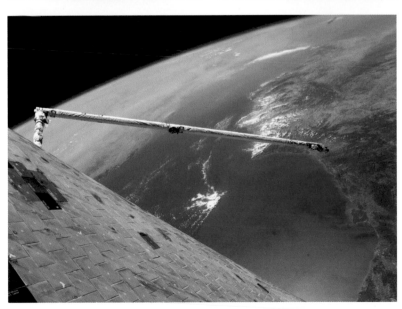

vital for protecting the Shuttle during re-entry. When *Columbia* was destroyed during re-entry on 1 February 2003, new ways were found to map any damage in areas not readily accessible to the crew. So a new Orbit Boom Sensor System (OBSS) assembly was provided for each Shuttle mission to enable the crew to gain visual access to the underside of the Orbiter, and every part of its surface, without having to go outside on an EVA.

Fabricated from two graphite epoxy cylinders, each 20ft long, they were originally manufactured as replacement arms for the Remote Manipulator System *(RMS, see page 160)*. The upper and lower booms are joined by a rigid fixture which can be held by the

end effector of the RMS. With special sensors attached to the other end, the articulating combination of RMS and OBSS allows the crew to visually inspect and to scan the entire exterior surface of the Orbiter after it has reached orbit. Despite all precautions, pieces of spray-on foam insulation used to insulate the External Tank can come loose during the dynamic vibrations of launch and ascent and other parts of the Shuttle, too, can break off and strike delicate tile or reinforced carbon-carbon wing leading edges or the nose cap.

As an example, on what was unknown to the general public because it was the third mission for the US Department of Defense, the second flight of *Atlantis* came close to being its last as a result of an incident during

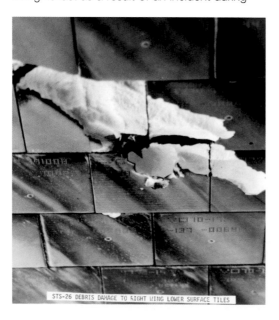

STS-26 DEBRIS DAMAGE TO RIGHT WING LOWER SURFACE TILES

ascent on 2 December 1988. It was the second mission after the loss of *Challenger* almost three years earlier and it was classified top secret because it was carrying a 40,000lb Lacrosse radar reconnaissance satellite. Even before *Challenger*, the military had backed away smartly from the Shuttle, seeing it as an expensive way to put satellites in space. But it had already built some satellites that could only be launched by the Shuttle, so a few still had to go aloft carried by NASA's winged Orbiter.

Shortly after *Atlantis* lifted off, a piece of an SRB nose cone became detached and apparently struck the side of the Orbiter, an event which became all too apparent when the video footage from high-speed cameras revealed the potential damage that could have been caused. The crew was asked to use the remote manipulator arm to inspect the area, using the camera to take pictures of the suspect area. What they saw was a portent of catastrophic disaster as one tile was completely missing and several score more were severely damaged. But because it was a military mission the crew had to download the video of the damage on encrypted signals, which seriously reduced the resolution of the images.

In Houston, engineers could not see the degree of damage visible to the crew on orbit, since they were watching the images on direct TV displays in the Orbiter. Several strong exchanges with the ground resulted in engineers overriding the concerns of the crew. During re-entry on 6 December 1988, on being exposed to the full heat of re-entry, a portion of the right side of the

forward fuselage began to burn through and the aluminium structure started to melt. The Orbiter was saved from fiery destruction only because the metal in that area was much thicker than elsewhere due to the fixtures for the Ku-bad antenna. On inspection after landing, more than 700 tiles had been affected by the impact – prescience for the day almost 14 years later when *Columbia* would burn up after a piece of foam shattered a section of her reinforced-carbon-carbon wing leading edge heat shield.

Alarm systems

Nobody likes surprises less than an astronaut in space, where a problem can develop into a crisis very quickly. One of the most important provisions for crew information is the caution and warning system (CWS), which is designed to warn the crew of malfunctions and irregularities that need the immediate attention of an astronaut. Hardware and electronics provide visual and audible

ABOVE LEFT The damage to *Atlantis* was caused by a SRB nose cap coming off during ascent and leaving a trail of damage that could have opened a thermal path and caused a full burn-through on re-entry. *(NASA)*

ABOVE Shamefully, the scale of damage to the thermal tiles failed to stimulate development of an OBSS system *(page 153)* until it was finally incorporated after the loss of *Columbia*. *(NASA)*

CENTRE Mission Control at the Johnson Space Center, outside Houston, Texas, from where US manned space flights have been controlled since 1965. *(NASA)*

LEFT Watching from space. A photograph of the Houston district and the Johnson Space Center from an orbiting Shuttle. *(NASA)*

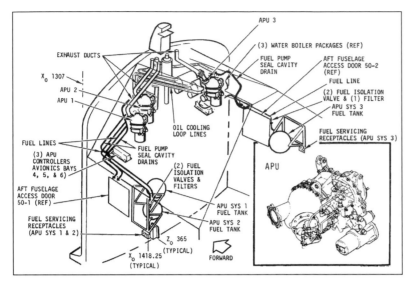

ABOVE The Auxiliary Power Unit (APU) provides the power to drive the hydraulic units for gimbal actuators on the three main engines, and for aerodynamic control surfaces and the landing gear during descent. *(NASA)*

Fire prevention

The threat of a fire inside the Shuttle Orbiter has potentially catastrophic consequences, although the mixed-gas, oxygen/nitrogen atmosphere at sea-level pressure is a lot better than the pure oxygen environment of previous US manned spacecraft. It was in one of those, during countdown rehearsals on 27 January 1967 for the first Apollo mission, that astronauts Grissom, White and Chafee lost their lives when an electrical short circuit caused a violent conflagration. But the fire had been in a 1g environment on earth and in a highly pressurised pure oxygen atmosphere where even a match would produce a flame four times its natural length. When Apollo first flew in October the following year, the spacecraft was virtually fireproof.

In weightlessness without convection fire behaves very differently and in an environment where oxygen is only one-fifth of the atmosphere, flame propagation rates are low. Because the flame remains glowing in a spherical form, it tends to snuff itself out, consuming the oxygen in its immediate vicinity without which it will not burn. But there is no compromise and the Orbiter has a Smoke Detection and Fire Suppression System (SDFSS) fitted in a crew cabin avionics bays,

cues that something is wrong. There are four master alarms – a 40-light array on one of the display panels, and a 120-light array on another. The audible cue goes to the flight crew headsets and to the speaker boxes. The cue sensors interface with the auxiliary power units, the data processing system, electrical power system, flight controls, guidance and navigation, hydraulics, the main propulsion, orbital manoeuvring and reaction control systems and the various payloads. The emergency alarms consist of a siren, which is activated by a smoke detection system and a klaxon which warns of a loss of cabin pressure.

RIGHT Flying the Shuttle takes place on the flight deck, while the living quarters are in the mid-deck. However, during rendezvous and approach manoeuvres, one of the two pilots will 'fly' the Orbiter by using the rotational hand-controller on the aft 'on-orbit' station (top-right drawing), where there are two aft and two upward facing windows. *(NASA)*

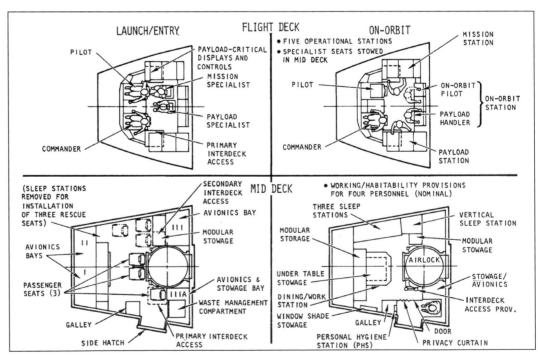

the crew cabin itself and in any pressurised modules in the cargo bay. The SDFSS consists of ionisation detectors located throughout the vehicle and will emit an alarm if they sense a smoke concentration of 2,200 microgrammes/ cu m, or a rate of smoke increase of 22 microgrammes/cu m/sec across eight consecutive counts.

A fixed fire suppressant bottle consisting of a 3.8lb Freon-1301 is built in to each of the three crew cabin avionics bays, activated via a control panel switch. Three hand-held fire extinguishers are in the crew compartment, one on the flight deck and two in the crew mid-deck area. Each bottle is 13in long and tapered to fit into any one of numerous fire suppression ports in the

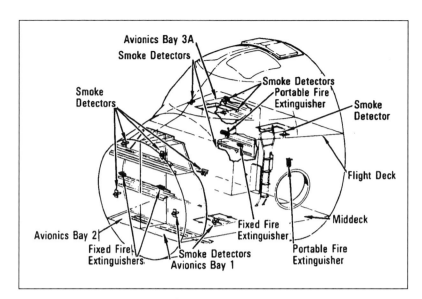

ABOVE Smoke detectors and fire extinguishers provide alert and response devices to deal with any fire that may break out, however unlikely, with particular emphasis on electrical and avionics equipment. (NASA)

LEFT Extinguishing ports for inserting the nozzles of the fire extinguishers are situated in a wide variety of locations around the pressure cabin. (NASA)

BELOW The aft work station, from where control of the Shuttle's rendezvous manoeuvres is conducted, and from where the remote manipulator arm is operated. (NASA)

RIGHT The two forward-facing pilots' seats remain fixed throughout the flight. Note the hand-controller for commanding manoeuvres and 'flying' the Orbiter after re-entry. *(NASA)*

ABOVE LEFT The Orbiter mission trainer provides a scale impression of the amount of room available in the mid-deck, with the ingress/egress hatch at the back and the airlock hatch at left. The forward lockers are to the right, behind which are the forward avionics bays. *(NASA)*

ABOVE In space, Shuttle astronauts peer through the aft-facing windows into the payload bay area. TV screens for operating the remote manipulator arm and laptops can be seen on the port wall. *(NASA)*

BELOW With launch and re-entry seats folded and stowed, there is more room to conduct experiments, in this case work known as Electrophoresis Operations in Space (EOS) on a potential vaccine to counter diabetes. *(NASA)*

RIGHT When docked to the International Space Station, access from the Orbiter is via the airlock module mounted in the forward payload bay. Before the ISS, the airlock module was usually carried in the mid-deck area. *(NASA)*

RIGHT *Atlantis* displays its cargo of modules for the International Space Station. The remote manipulator arm is attached to the port side of the payload bay. *(NASA)*

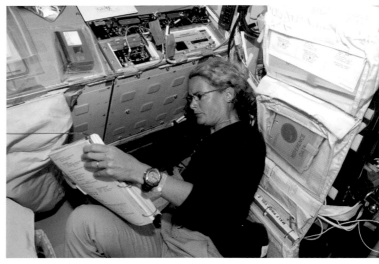

ABOVE Where there is no 'up' or 'down', orientation is selective and can lead to vestibular disorientation (confused perception of direction). About 50 per cent of astronauts become space sick for a few days, before vomiting declines. *(NASA)*

ABOVE RIGHT Tucked down behind the commander's seat on the flight deck of *Endeavour*, Canadian astronaut Julie Payette goes over checklists and books from the Flight Data File during STS-127. *(NASA)*

LEFT Crewmembers from STS-128 revise assembly manuals for equipment brought up to the International Space Station. *(NASA)*

FAR LEFT Karen Nyberg has fun with the camera as she suspends herself weightless in the mid-deck area of *Discovery* during STS-124. *(NASA)*

LEFT The crew of STS-124 mixes with astronauts aboard the International Space Station for a communal meal. *(NASA)*

displays and control panels. It contains Halon-1301 which suppresses the effects of the fire as well as acting to extinguish combustion, and can also be used as back-up in the avionics bays.

Lighting

Lighting the Orbiter, inside and out, must take account of the varying degree of ambient illumination from natural and artificial sources as the spacecraft moves in and out of night once every 90 minutes on orbit. The interior lighting provides for displays and controls as well as crew equipment, while exterior lighting illuminates payload bay door operations, EVA, operations with the remote manipulator system (RMS) and for station-keeping and docking. Floodlights are controlled from the crew compartment and consist of metal halide lamps similar to mercury vapour lamps. Inside the crew compartment, lighting falls into three categories: panel lighting, which helps the crew see nomenclature and displays; instrument lighting, incandescent lamps behind the instrument face for reading displayed data; and numeric lighting, six indicators on the flight deck numeric readouts to display data.

RIGHT The remote manipulator arm has picked up the OBSS *(see page 153)* **and is providing the crew inside** *Discovery* **with a sensor scan of the exterior surface, checking for damaged thermal insulation tiles.** *(NASA)*

BELOW When the manipulator arm was built by Canada's Spar Aerospace, it could manoeuvre loads of up to 65,000lb. With upgrades and improvements over the last 30 years it can now handle a mass of up to 586,000lb without the gears slipping. *(NASA)*

Remote Manipulator System (RMS)

The only piece of equipment routinely carried by the Orbiter and not built in the United States, the Remote Manipulator System (RMS) is the mechanical arm that moves payloads in and out of the cargo bay and grasps equipment, satellites or spacecraft to hold in a fixed position so that astronauts can work on them. It must also support an astronaut on EVA, which the RMS can lift into position alongside a separate structure such as a part of the space station, enabling work to be carried out from a platform fixed to the end of the manipulator arm.

Designed and developed by Spar Aerospace of Canada, it is very big, about the size of two telegraph poles laid end-to-end, and can manipulate the maximum mass capable of being carried in the payload bay – around 30 tons – to within a fraction of an inch of the desired location. Canada has always had a strong aerospace industry and during the late 1950s and early 1960s many of Canada's finest engineers and scientists headed south to cross the border, join NASA and work on many US space programmes, at the same time as many young

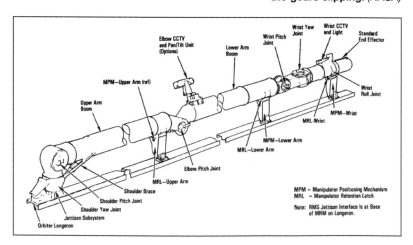

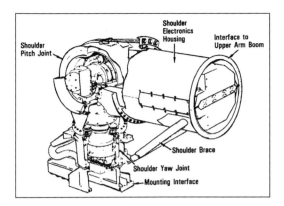

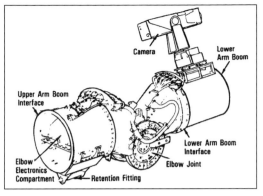

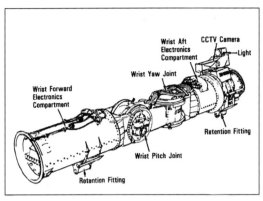

FAR LEFT The Remote Manipulator System (RMS) is attached to the Orbiter at the forward end of the payload bay on the port side, a position known as the shoulder joint, which can move in pitch and yaw. *(NASA)*

LEFT The reinforced graphite upper and lower arm sections are articulated at the elbow joint and can move in pitch. *(NASA)*

FAR LEFT The wrist joint can move in pitch, roll and yaw, and connects the lower arm to the end effector. *(NASA)*

LEFT The end effector grips the probe on a grapple fixture using snare wires that pull tight to create a rigid connection. *(NASA)*

students and engineers were heading west from Europe to do the same – this author included!

The design of the RMS copies the articulated segments of the human arm. The shoulder joint is where the RMS is fixed to the forward end of a fuselage longeron forming the left sill of the payload bay, inboard of the hinge carrying one side of the payload bay doors. The elbow joint is halfway along the arm, which terminates in a wrist joint. The function of a human hand is performed by an 'end effector' which is used to grapple the object to be manipulated. The total arm is 50ft 3in long with a diameter of 1ft 3in and a total weight of 905lb, or 994lb with all the associated equipment needed to support and operate it.

The arm has a total of six articulating joints: yaw and pitch joints at the shoulder, a pitch joint at the elbow and pitch, roll and yaw joints in the wrist. Just like a human arm, upper and lower arms connect the three clusters of joints, 17ft and 20ft long respectively, and made from graphite epoxy for low weight and high strength. The standard end effector can be used with a wide range of payloads and can carry electrical connections through to the object being grappled, providing power from

the Orbiter to the payload. Special purpose-built end effectors can be used from time to time, depending upon the specific requirements of a particular mission.

Operation of the RMS is conducted from the aft flight deck, utilising a combination of window views and TV images from CCTVs usually installed on the elbow, with pan and tilt capability, and the wrist, which also carries a light. Four CCTV cameras in the payload bay can also be used to aid in manoeuvring the RMS and these each have pan, tilt and zoom capability. The two CCTV monitors at the flight deck station can each display any two TV cameras in split-screen mode, allowing the astronaut to work with four different views. The tip of an unloaded arm cannot be moved faster than 2ft/sec (about 1.4mph) and a maximum load cannot move faster than 0.2ft/sec. Yet for its slow motion, it is incredibly accurate, and able to position a peg within a 0.06in hole.

Although weightless, in space loads have inertia because the mass is the same whether on earth or in orbit. The weight is merely a product of the gravitational attraction exerted by a major body (planet) on an object of fixed mass. On the moon, an object will weigh one-

Communications

As ET found out, phoning home is not always easy and keeping in touch with Mission Control in Houston is a global problem when the Shuttle orbits the earth around 16 times a day. The problem is magnified by the fact that the earth spins one complete revolution every 24 hours, so the ground track over which the Orbiter flies is constantly migrating west as the planet (viewed from above the North Pole) spins anti-clockwise. This means there is no single track around the earth along which to set up tracking and communication stations and this has been a problem since the dawn of the space age.

Before the Shuttle, various tracking stations were scattered around the earth, some in foreign countries, some on ships at sea and sometimes aboard aircraft. Big gaps in coverage, waiting for the spacecraft to pass within range, was unavoidable. But that has changed and more than 80 per cent of the Shuttle's ground track is covered. This is made possible through the Tracking and Data Relay Satellite (TDRS, pronounced 'tee-driss') system which involves two powerful communication satellites placed in geostationary orbit 22,300 miles above the earth. At that distance, with the satellites aligned with the equator, they take 24 hours to orbit the earth and so appear stationary over one spot on the planet's equator.

Because TDRS satellites are at such a

ABOVE Complete with life-support backpack, spacewalking astronauts are tethered to the Orbiter by a thin wire attached to a running line along the side of the payload bay, just discernible at extreme left. *(NASA)*

sixth the weight it has on earth; on Mars, one-third. But the inertia of a fixed mass is the same everywhere. Even in weightlessness, an object sent at speed into the side of a solid structure will do just as much damage as it would on the surface of the earth. This is why the careful manipulation of large objects in space is vital for reasons of safety. This is why a series of precautionary interlocks in the controlling software prevent inadvertent use of the RMS in a dangerous manner.

RIGHT

Communications through the TDRS satellites has considerably improved the amount of orbit time during which the crew are in contact with the ground, and the volume of data that can flow between the Orbiter and Mission Control. *(NASA)*

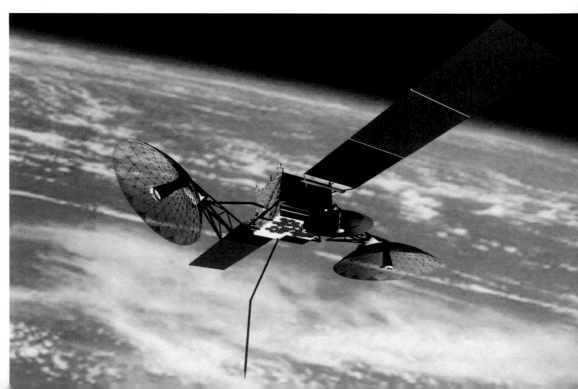

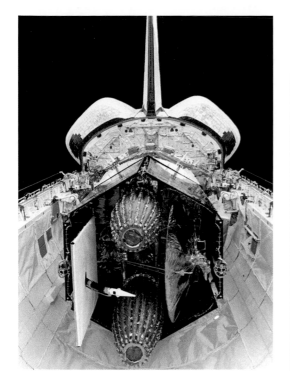

FAR LEFT The
TDRS-C satellite in the
Orbiter payload bay,
from where it will be
deployed to low earth
orbit from STS-26, the
first mission after the
loss of *Challenger*.
(NASA)

LEFT TDRS was a
solution to the problem
of orbital coverage
over large areas of the
earth's surface, with a
series of such platforms
also routing data
and communications
from a wide variety of
unmanned satellites.
(NASA)

great distance (almost three times the diameter of the earth) they are within line-of-sight of everything that passes across almost half of the earth's surface. The Shuttle is orbiting at a height of only 200–400 miles and so TDRS is in a position to relay signals between the Orbiter and Mission Control in Houston. By transmitting all the way out to a TDRS satellite, the Shuttle can stay in touch very much longer than if it relied on flying over ground stations. With the TDRS satellites spaced apart, they cover most of the world and are used to relay, collect, gather and re-transmit data from a wide range of satellites, negating the need to maintain a global land-based network.

TDRS is used for S-band phase modulated communications between the Orbiter and the ground most of the time, the uplink signal being on a frequency of 2,106.4MHz or 2,041.9MHz, thereby allowing communication with two Orbiters if ever needed. The phase modulated downlink signal is on a carrier wave of either 2,287.5MHz or 2,217.5MHx. Both uplink and downlink frequencies can carry voice, data or a combination of the two at 72kbps and 192kbps, respectively. The S-band frequency modulated carrier wave cannot receive information from the ground, but can transmit from eight different sources. Two S-band phase modulated antenna quadrants are attached to the forward fuselage and immediately aft of the windows but forward of the wing leading edge, covered by thermal insulation tiles.

For use only on orbit when the payload bay doors are open, the Shuttle also has a graphite epoxy Ku-band antenna, 7ft long and with a deployed diameter of 3ft, weighing 304lb. It is mounted to the right forward sill longeron, looking forward along the cargo bay, immediately to the rear of the forward mid-fuselage bulkhead. It operates on the Ku-band frequencies of 15.250GHz and 17.250GHz, with a TDRS carrier downlink frequency of 13.755GHz and 15.003GHz from the Orbiter to the TDRS. It has a full gimbal capability in two axes and can downlink data at 50mbps or transmit through the TDRS at 216kbps. Because the Ku-band system has a narrow beam width, with a much wider beam width the S-band signal is used first to lock the Ku system into position. The Ku-band system can also be used as a radar device for skin tracking or for triggering a receptive transponder and has a 60° scan arc, but it cannot be used for both purposes simultaneously.

A third tier of communication is the humble UHF system, essentially operating a back-up to the S-band and Ku-band systems. It is

used for communications with the Spaceflight Tracking and Data Network (STDN), the legacy of NASA's old ground-based system. Operating in simplex mode, it cannot transmit and receive simultaneously. Two frequencies, 296.8MHz and 259.7MHz, are available for both transmitting and receiving, while 243.0MHz is also available for receiving. UHF frequencies are also available for EVA, when two astronauts are outside the Orbiter.

Voice and data from one astronaut operating in Mode A will go to the Orbiter on 259.7MHz, while simultaneously transmitting voice only to the second astronaut on that frequency too, receiving voice from the Orbiter on 296.8MHz

and from the second EVA astronaut (on Mode B) on 279.0MHz. The second astronaut on Mode B transmits both data and voice to the Orbiter on 279.0MHz, transmitting voice only to the other astronaut on the same frequency. The Mode B astronaut receives voice from the Orbiter on 296.8MHz and from the other EVA astronaut on 259.7MHz. The astronauts in the Orbiter switch the UHF configuration to send this communication in relay mode to the S-band system and down to the STDN stations, via TDRS or by Ku-band. Biomedical data is separated from voice signals and transmitted to the ground.

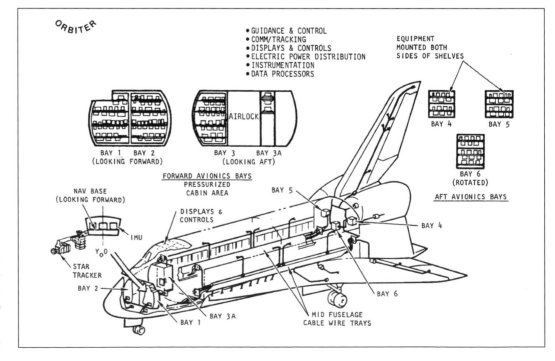

Avionics

With more than 300 separate pieces of equipment located around the Orbiter, the avionics system is easily the most prolific and, arguably, the most important. So important in today's aerospace vehicles, avionics are by far the greatest cost in the price of a new combat aircraft, sometimes reaching as much as 70 per cent of the purchase price – the airframe and the engines accounting for little more than one-third of the cost. This is because the technology to make the most of aerodynamic shapes that look strange but fly right is embedded within the flight control systems that operate faster and with much more sensory input than any human ever could. And doing that is expensive.

When the Shuttle avionics were developed in the early 1970s, much that it would rely on was new. Solid state electronics was not all that old and the transition to fully automated systems was bridged by the Shuttle. Developments in processing power and memory storage had been so great that the Hewlett Packard calculator strapped to astronaut John Young's knee when he commanded the first Shuttle mission in 1981 had even greater processing power than that carried in Apollo during the moon landings of just a decade earlier.

Without the avionics system the Shuttle could not fly and the heart of all its operations, involving the External Tank and the Solid Rocket Boosters as well, lies within the 300 miles of wiring inside the Orbiter, linking through 6,500 connectors to more than three tons of computers, guidance and navigation equipment and a host of subsystems and electronic equipment to control and navigate the Shuttle as both spacecraft and aircraft. With continuous upgrades and improvements that keep track of the latest developments, not all the desirable modifications engineers would like to install have found their way into the Shuttle. But enough have done so to make it matter.

The main data processing system comprises five general purpose computers (GPCs), two magnetic tape mass memory units each with 134-megabit bulk storage capability, modified standard data bus (Mil-Std-1553) plus a host of multiplexer/de multiplexer units for the Orbiter, the main engines and the SRBs, in addition to events controllers, timing units and data isolation amplifiers. Four of the GPCs are situated in the forward mid-deck avionics bay with the fifth in the aft mid-deck bay. Usually no longer installed, a sixth GPC was carried, disconnected in the forward avionics bay, just in case the others failed or suffered a universal bug.

During the first decade of Shuttle operations

LEFT Astronauts James Wetherbee and pilot Paul Lockart on the flight deck of *Endeavour* during the STS-113 mission to the International Space Station in November 2002. The Orbiter carries more than 2,100 display and control items on the flight deck, with head-up displays fitted to *Challenger*, *Discovery*, *Atlantis* and *Endeavour* during manufacture and retrofitted to *Columbia*. (NASA)

the initial IBM 4Pi/AP-101B computers were each equipped with 106,496 32-bit words of random access non-volatile memory on ferrite core, because nobody knew for sure the effects of solar radiation on solid-state RAM. Updated computers were developed by IBM through the 1980s and from early in the new decade the AP-101S began to replace the original model. This evolved from a computer originally designed for the Rockwell B-1B bomber, but it too was morphed into a hybrid, taking the same central processing unit and connecting it to a new 256k 32-bit solid-state hardened memory. This allows the new software to operate with the same sets of instructions developed for the

AP-101B. The AP-101S is highly reliable, with a Mean-Time-Between-Failure (MTBF) of around 6,000 hours, much better than the 1,000 hours factored in to the contractual requirement when the Shuttle programme began.

One of the most important groups of in-flight operations is to do with guidance, navigation and control (GNC) and this is governed by four of the five GPCs. They operate together, checking each other's calculations and serve as a redundant set. It is the core requirement in the Shuttle system, first set when the Shuttle itself was a concept that had yet to be designed *(see Chapter 1)*, to be able to fail-operational/ fail-operational/fail-safe. A failure in any one computer will not imperil the function because there will be four others to vote against its error and maintain operability. Core to its operation is the Primary Avionics Software System (PASS) which is arguably the most advanced and sophisticated software system ever produced for an aerospace programme.

Software for the data processing system is divided into separate categories covering systems and applications, which together form the memory software for a particular flight. The system software is loaded initially into the five GPCs and controls operations between the computers, maintains time and a host of standard activities. The applications software is there to do all the jobs necessary to fly

the Shuttle and conduct on-orbit operations. In addition to supporting all launch, ascent, manoeuvring, re-entry and landing operations via the GNC, it also conducts systems management duties and payload functions. Systems management is all about Orbiter life support, thermal control, communications and payload operations and is run by a single GPC in simplex memory configuration. The payload functions are largely unused in the form for which they were designed and interact with ground preparation of the vehicle for flight.

Flight operations are divided up into a series of duties embraced by operational sequencers, known as OPS. Each segment of the mission contains its own batch of software and each is numbered according to function. The crew designates which batch of software will be uploaded to a GPC from the mass memory unit according to upcoming activities. These OPS functions are further divided into major modes (MM) so that each can contain several MMs operating in sequence. While all the relevant sub-functions of an OPS can be loaded as MMs into the GPC, only a single OPS will usually reside in the computers. One exception is the OPS 1 (ascent) software which will also be loaded along with OPS 6 (abort) software, since the latter would need to be performed quickly and without the time for a crewmember to load the programme.

From the first Shuttle flights, displaying information to the flight crew was made through the Multifunction CRT Display System (MCDS) comprising four display units including cathode ray tubes, three keyboard units for communicating with the main general purpose computers and controls for brightness levels and general presentation data. With its CRTs and electronic displays the Shuttle was in advance of its time, but during the following decade so many developments took place in cockpit information and display technology that it quickly became out of date. To begin with, neither *Enterprise* nor *Columbia* had head-up displays (HUDs) which were soon to become commonplace in aircraft. *Columbia* did get HUDs installed during its stand-down in 1984–85 when major upgrades and modifications were installed.

By the early 1990s, Rockwell had made progress with a new Multifunction Electronic

Display System (MEDS), incorporating all the latest technology for a fully glass cockpit, bringing the Shuttle up to date with the latest design configurations in the commercial airliner world. The MEDS has eleven identical colour display units, four integrated display units and four analogue–digital converters. Nine of the display units are installed in the forward flight station and two are in the aft station. The main display units derive from those developed and installed on the Boeing 777 airliner, each screen being 6.7in square with LCD screen resolution of 99.3p/in. The MEDS was first installed on *Atlantis* in 1998 and first flown into space in May 2000, followed by *Columbia*, *Endeavour* and *Discovery*.

Chapter Six

Living on the Shuttle

The Shuttle has been a stepping stone to the International Space Station, providing the heavy lifting capability to build the world's largest research facility in orbit. It has also pioneered the way ordinary people can live and work in space. Compared to all previous spacecraft, the Shuttle provides astronauts with spacious living accommodation, and future generations of astronauts will be in debt to the Shuttle for this alone. Nevertheless, separation from loved ones on earth can bring great pressure to those faced with life-threatening dangers in space.

LEFT July 2009, and nine of 13 astronauts on board gather in the International Space Station to eat and chat about the ordinary things people talk about when meeting up. As one crew member told the author; "You can take the people out of the Earth but you cannot take the Earth out of the people!" *(NASA)*

Managing weightlessness

In weightlessness there is no up or down and objects, humans included, tend to drift around at random. This lack of anchor can be both inconvenient and hazardous when conducting switch changes at panels around the crew compartment. In Apollo, space was so cramped it mattered little but the interior

volume of the Shuttle is very much greater and the crew has a wide range of waist tethers, foot restraints and aids to keep them where they want to be, their hands free to operate switches and controls rather than hold themselves in one place. This is essential when hands are needed to stop things floating out of reach, particularly manuals and books. One such collection is the flight data file, or FDF, which is a collection of documentation, maps, procedures manuals, cue cards, star charts, earth maps, notebooks and writing materials. It is the equivalent of the airline pilot's flight bag and is kept in containers to left and right of the pilot seats on the flight deck and in two lockers in the mid-deck.

Daily routines

The 24-hour 'day' usually provides an 8hr sleep period flanked at each end by a 45min period for personal hygiene. Each crewmember has a kit for brushing teeth, hair care, shaving, nail care and appropriate health care items for female astronauts. Two fresh washcloths and one fresh towel are provided for each astronaut every day but absorbent, multiply, tissues are available from two dispensers per crew member for each period of seven days, lint free, of course, to prevent particles floating into the filters. Meal periods punctuate the working day but assignments are divided up on a flight plan that provides the crew with assigned tasks throughout the work hours.

Housekeeping duties are divided among all the crew, each requiring individual astronauts to spend 5–15 minutes at various times in any 24hr period cleaning the waste management compartment, the dining area, floors and walls, cabin air filters, trash collection and disposal and changing the lithium hydroxide canisters that remove excess carbon dioxide exhaled by the crew. Cleaning materials, rubber gloves, wipes and a vacuum cleaner ensure that proper cleanliness reduces to a minimum the chance for bacteria to grow in the confines of the crew compartment. Trash containers are located at various places with airtight seals to prevent germ contamination but if the bags begin to inflate due to biological activity a 41in vent hose is attached to an access port and the gases vented overboard.

Cameras

Cameras are carried by the crew to document activity inside and outside the Orbiter and include 16mm, 35mm and 70mm equipment. The 16mm camera is like a small motion picture camera with independent shutter speeds and frame rates. The 35mm Nikon is motorised with reflex viewing and, standard manual and three automatic modes for single or continuous exposure using an f/1.4 lens. The 70mm single-lens reflex automatic Hasselblad is a modified battery powered motor camera and comes with optional 80mm and 250mm lenses and film magazines, each containing 80 exposures.

Eating in space

Food trays for either the warmer or the full-scale galley are also stored in the mid-deck lockers, specially colour-coded for each astronaut with Velcro along the bottom for attaching to the locker surface and with straps for attaching to the astronaut's knees during mealtimes. Cutouts on the tray hold pots and cups much like drink-grips in cars. Two magnetic strips hold the utensils with condiments, gum, candy, vitamin tablets, wet wipes, dry wipes, drinking straws available as necessary.

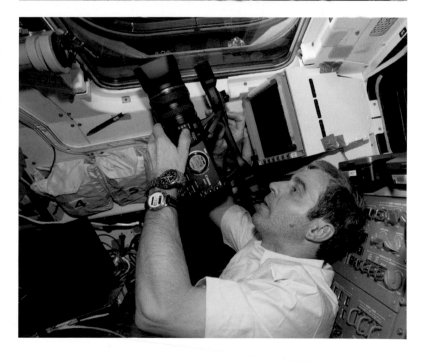

With weightlessness dominating the design of everything to do with food and eating, and where there is no conventional up or down, food will remain where it is even if the tray is rotated. Canisters, pouches, cups and anything designed to contain food has been designed with that in mind. While the consistency of the food is very close to that eaten on earth, there is a tendency for everything to be contained beneath lids or covers. Meals prepared for the airline industry have to take account of changes to passengers' taste buds at high altitude. Similarly, in space, everything tastes different and preparation of meals on the ground for packaging aboard the Shuttle must take that into account too.

Food

The food was divided into menu food or pantry food. Menu food was provided for the three set eating periods and was selected to give each crewmember 2,700 calories per day. They were divided into different storage types, including fresh, heat-stabilised, rehydratable, irradiated, moist and natural-form foods. Two days of emergency rations were added as pantry food, with 2,100 calories per day, but it also provided for snacks and beverages between meals.

Fresh fruits (consisting of apples, oranges and bananas), vegetable sticks, bread, breakfast rolls and cheese, stored either without wrappers, in zip-lock plastic bags or in the original manufacturer's wrapper (bread being an obvious one). Heat-stabilised food came in cans or foil-backed plastic pouches in which

the contents were heated before eating. They included cheese spread, beef, tuna, etc. Items such as scrambled egg, beef patty, chicken and noodles stored in plastic containers could be rehydrated with water. Similarly, rehydratable beverages such as coffee, lemonade or orange drink would be turned liquid by water injection and taken down via a small plastic tube inserted through a cap. Some foods such as meats would have been irradiated and held in a plastic pouch and cut open with scissors. Ready foods such as nuts would come in a plastic pouch much like those on sale in shops and supermarkets.

In all, a three-meal-per-day food selection would be found in about 20 packages balanced with 4–7 rehydratables, 1–4 heat stabilised or irradiated types, 8 beverages and 4 ready meals. The art was in the preparation of the food but science was at the base of the dieticians' menu board and detailed preparation of the food selected for a particular flight could easily involve nearly 300 separate meals. With each crew member allocated approximately 280 packages for a 14-day flight, the pantry could easily be called upon to store nearly 2,000 items for all seven astronauts.

Food preparation

Equipment to prepare the food comprises a flexible line 10ft long for supplying water at ambient temperature or chilled, with a special rehydratable food container station where it can be reconstituted, mixed or heated. Likewise, the rehydratable beverage containers have a lip with orifice for inserting a water spigot. A portable food warmer is flown on some missions instead of a galley, particularly when the Shuttle is visiting the International Space Station and astronauts will be using the mid-deck food preparation area for only two days before docking and two days prior to landing. Food is heated through thermal conduction on to a hot plate because in weightlessness there is no convection. Hot air will not rise, nor will cold air sink. An aluminium case, 13in x 18in x 6in in size, the food warmer looks like a small attaché case within which are placed the items to be heated. It is attached to a power cable 12ft 10in long and can warm a meal for four crewmembers in one hour. Up to 14 rehydratable packets can be handled simultaneously.

The galley is carried on flights where the Shuttle is not docking to the space station and is a multipurpose facility located on the mid-deck where a single nominated crewmember can prepare all the food for a single meal for all crewmembers. It is divided into separate compartments for rehydrating and warming a variety of food pouches. This is a food preparation station in that it provides everything necessary for cooking up a full crew meal, with water spigots, condiment dispensers, wipes, lights for indicating cooking status and provision for attaching containers and pouches identical to those used in the food warmer.

Sleeping on the Shuttle

Getting sleep aboard the Shuttle depends on the mission being flown. On a single-shift flight where all the crew sleep at the same time, various sleeping bags, tethers and restraints are carried and set up inside the mid-deck area. Weightlessness produces some strange effects. Unrestrained arms float up above the chest and hang in front giving astronauts the appearance of nocturnal zombies. Rigid sleep stations are carried on missions where work will continue around the clock, dividing the crew into those at rest and those on duty. Two types are available, providing cubicles for three or four crewmembers, places where they can get some isolation and a degree of privacy while others work around them outside the panels of the cubicle. When vacated, the working group settles down to rest, after a meal together around the mid-deck.

Medical matters

Serious illness in space would result in the Shuttle returning to earth but an onboard medical kit comprising two packages is available for a host of minor illnesses, injuries and ailments, from toothache to open wounds. The two medical packages are stored in the mid-deck forward locker, one coloured blue contains bandages and medications, the other is coloured red and contains emergency equipment. A

wide range of tablets is available including anti-histamines, anaesthetics, anti-inflammatories, tranquillisers and tablets for motion sickness and nausea. About half of all astronauts get space sickness, which is similar to motion sickness and sea sickness on earth, and usually lasts about three days. The cause too is the same, a disruption in the vestibular function of the otolith in the ear giving disruptive signals to the body's balance mechanism. The treatment is usually a Scop/Dex or Phenergen suppository.

Numerous items of medical equipment are on hand including syringes, antiseptic wipes, urine test packets to determine protein, sugar, blood and pH levels, tweezers, scissors, scalpels, sutures, tape and bandages. Blood pressure measurements can be taken, oral airways examined, an opthalmoscope can be used to examine eyes, and microbiological test kits could provide information about cultures from throat, urine or wounds. If an astronaut should need monitoring due to some cardiac problem, bioinstrumentation is available with electrodes attached directly to the skin and plugged in to the Orbiter's telemetry system for transmission to earth. Astronauts enjoy the same patient/doctor confidentiality as they would on earth and private communication sessions can be arranged. Routine checks are part of a standard mission and radiation monitoring is a routine activity. Crewmembers carry passive dosimeters similar to those worn by employees in the nuclear power industry and these are also placed at various locations around the crew compartment.

ABOVE Astronaut Scott Kelly was aboard the STS-118 mission that carried school teacher Barbara Morgan into space on *Endeavour* in 2007. He was also on the International Space Station when his twin brother's wife, Congresswoman Gabrielle Gifford, was shot and wounded on 8 January, 2011. His brother, Mark Kelly, is also an astronaut, assigned to command the penultimate Shuttle mission, OV-105 *Endeavour* on STS-134. Living in space, separated in times of crisis, can be one of the hardest tasks for those left waiting. (NASA)

Chapter Seven

Coming home

The Shuttle is the only vehicle to have carried astronauts on flights into space and to have returned them to land on a conventional runway. Even so, the challenge of surviving the tremendous heat generated by friction with the Earth's atmosphere on re-entry brought challenges to Shuttle engineers long before it first flew – and placed great demands on those astronauts who made the first pioneering Shuttle flights into space. The tragic loss of *Columbia* on 1 February 2003 demonstrated that re-entry will never become routine.

LEFT *Discovery* touches down at Edwards Air Force Base, California, after returning from its mission to the International Space Station in 2009. Note how the No.2 and No.3 main engines have been gimbaled up to minimise the effect of heating. *(NASA)*

ABOVE An electron beam is used to measure the thermal flow on the aerodynamic shape of the Orbiter during simulated re-entry in a laboratory. *(NASA)*

RIGHT The Flight Director Attitude Indicator (FDAI) provides attitude and rate information on the Orbiter as it descends after re-entry. It is a development of a standard flightdeck instrument familiar to all pilots. *(NASA)*

FAR RIGHT The Reinforced Carbon-Carbon nose cap visibly displays the thermal effects of temperatures reaching almost 3,000°F during re-entry. *(NASA)*

During orbital flight the Shuttle is a spacecraft operating on systems developed from similar technology pioneered on earlier manned spacecraft, but when it comes to return home it morphs itself into an aircraft and flies back down to a runway – unique among all piloted vehicles that have ventured into the atmosphere and beyond. It is this that makes the Orbiter so special – first of a kind and to the very end of its operational life, one of a kind.

Preparation for re-entry begins a day before the event, with the crew stowing everything in its assigned location. Having been weightless for some considerable time, it is easy to forget that when the vehicle starts to descend through the atmosphere and picks up the first sensations of gravity, anything not properly secured will be thrown around the crew

compartment causing damage or injury to the crewmembers. More than just an exercise in military-style neatness, putting everything away securely is a matter of getting back safely. The Digital Autopilot must be loaded with appropriate software and all the systems checked, some of which will not have been operating since shortly after reaching orbit.

About four hours prior to re-entry, the environmental control system and life support radiator bypass/flash evaporator must be checked. It alone will be used to cool the Freon-21 coolant loops after the payload bay doors are shut and the radiators no longer operable. The evaporator will be used until the ammonia boilers are brought on at around 140,000ft altitude. After preparation of the coolant system the inertial measurement units in the guidance and navigation system are aligned and the flight software (OPS 3) is loaded into the four main guidance computers.

The RCS thrusters orientate the Orbiter so that it is flying tail-first along the plane of the ground track, with the nose pitched down slightly so that the rocket motors in the OMS pods on the aft fuselage fire against the direction of flight, braking orbital speed by 250–450ft/sec (170–307mph) depending upon altitude. The Shuttle usually operates at an altitude of between 130 and 500 miles. The magnitude of the burn is determined by the orbital velocity and the reduction in speed necessary for its orbital motion to be overwhelmed by gravity, starting it on a shallow curve towards the atmosphere.

Before it gets there, the RCS thrusters

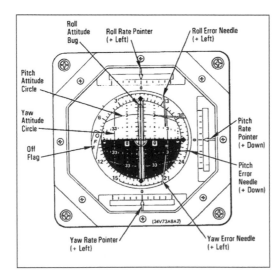

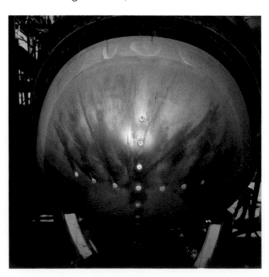

RIGHT From whichever direction, and whatever flight path azimuth the Orbiter approaches the landing site, it intersects imaginary circles in the sky that predict the degree of turn required to achieve the correct flight path for landing. *(NASA)*

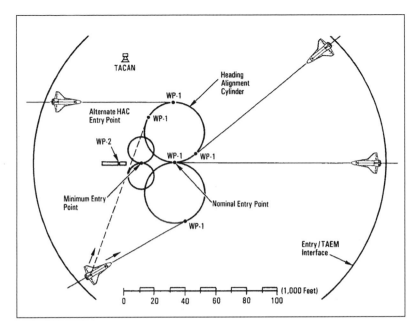

again turn the Orbiter around 180°, so that it is facing forward and pitched up about 40°. This presents a nose-up attitude so that the broad underside of the fuselage and delta wing slice into the atmosphere, which is first noticed by a slight deceleration at an altitude of 400,000ft (75.8 miles) and a speed of about 17,500mph. At that point, known as Entry-Interface (E-I), the Orbiter still has 5,000 miles to 'fly' as it descends through the atmosphere.

Re-entry

Entry itself is defined in three separate phases, because each has a unique set of software to control the Orbiter. Entry begins at five minutes before E-I and extends to an altitude of 83,000ft and a velocity of 2,500ft/sec (1,705mph) just 59 miles from the runway. During this phase the FCS will issue commands to the roll, pitch and yaw thrusters to hold attitude or change axial alignment for prescribed manoeuvres. The Orbiter is held to zero roll and yaw but with a pitch of 40° until E-I, where 0.176g is sensed corresponding to a dynamic pressure of 10lb/sq ft, at which point the ailerons 'feel' sufficient air pressure for them to be effective and the aft roll jets are deactivated. When a pressure of 20lb/sq ft is sensed the Orbiter's elevons become effective and the aft pitch jets are deactivated. The speed brake can be used below Mach 10 (ten times the speed of sound) for a positive elevator trim deflection, and below Mach 3.5 the rudder is activated and the aft yaw thrusters are deactivated, by which time the Orbiter will be down at around 45,000ft.

Long before that, at 265,000ft the Orbiter enters a communications blackout phase caused by ionising atoms of air around the vehicle, the Orbiter only emerging from this at about 162,000ft. The large amount of energy carried by the Orbiter as it re-enters the atmosphere must be dissipated and that could be achieved by one of two means or

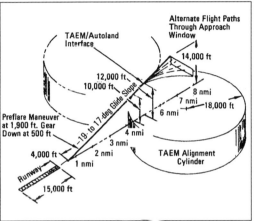

LEFT Seen in perspective, the heading alignment cylinders put the Orbiter into the terminal approach phase, where the descent rate is reduced in the pre-flare manoeuvre. *(NASA)*

BELOW Descending down the steep glide slope, the Orbiter pulls up at around 1,750ft to enter the shallow glide slope and the flare point. *(NASA)*

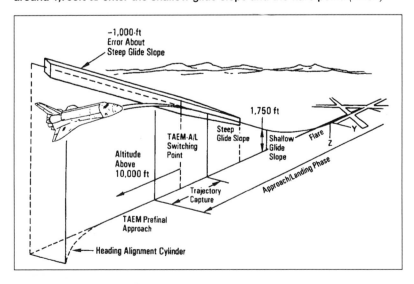

RIGHT *Atlantis* floats down from the flare to a perfect touchdown at NASA's Kennedy Space Center, tiny puffs of smoke signalling wheels down. The parachute helps to reduce loads on the brakes. Immediately after wheels-stop, the Orbiter is surrounded by vehicles designed to prevent fumes and fluids leaking out of the vehicle and posing a hazard to personnel on the runway and crew deplaning from the Shuttle. Toxic chemicals and hypergolic propellants must be removed quickly, and the vehicle checked for leaks so that it can be placed in a 'safed' condition. It can then be slowly towed to the Orbiter Processing Facility. *(NASA)*

combinations of both: it can use a steep re-entry trajectory to get rid of that through higher atmospheric drag, which will dissipate the energy at a much faster rate; or it can bleed that energy away through a series of roll manoeuvres, or high bank angles. Ideally, higher pitch rates would steepen the descent path but apply much higher heat rates, although the overall heat load would remain much the same. The difference is more than subtle.

If the Orbiter pitches up and presents more of the flat-iron effect to the atmosphere it will rapidly increase the friction coefficient and raise the temperature produced from heating the atmosphere through the discharge of kinetic energy. But doing that would significantly shorten the time it spends in getting down to the lower atmosphere and slower speeds where heat through friction is no longer a problem. It would, however, increase the deceleration g-force, which would require a stronger airframe and a tougher crew! For a shallower descent the peak temperatures are much less but the time spent getting down is significantly longer so that the total amount of heat loaded into the thermal protection system will be about the same as that accumulated by a steep descent taking less time.

The control logic maintains the Orbiter at the 40° pitch angle but rolls the Shuttle according to the amount of energy it has to dissipate, sometimes up to 70° bank angles being experienced. The roll angle decreases vertical lift and dumps energy which, while raising the temperature a little, is not as bad as using steeper descent angles and higher drag coefficients. The balance between lift and drag is a function of the distance the Orbiter has to fly to the runway, a term known as range-to-go. If the Orbiter is slightly short on distance it must dump greater amounts of energy, but if long it must impose less drag and extend its range. All the while, throughout this phase until speed is down to around 13,000mph, the heat rate must be maintained at a more or less constant level.

Within the roll component of the Orbiter's manoeuvres the computers must include the amount of cross-range it has to fly to reach the runway. Orbital ground tracks are fixed and will not necessarily fly directly over the landing site, so a degree of adjustment in the flight path will be necessary, the tortuous cross-range that

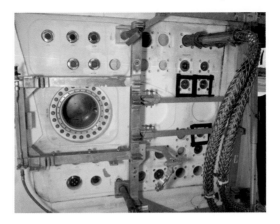

so greatly influenced selection of the Orbiter configuration (see Chapter 2). The equilibrium glide phase takes over from the temperature control phase and provides the maximum downrange glide capability, and is maintained until drag reaches a deceleration level of 33ft/sec/sq where the constant drag phase starts and pitch is gradually lowered to 36°.

Transition phase

The next, or transition, phase, begins when the Orbiter starts pitching down to 14° and an altitude of 83,000ft and a speed of 1,700mph, just 59 miles from the runway where it begins the Terminal Area Energy Management (TAEM) phase. This is where the guidance steers the Orbiter to the nearest of two imaginary cylinders in the sky, each with a radius of 18,000ft at a tangent to each other and on either side of the runway. These heading alignment cylinders (HACs) provide options for cases where the Orbiter has excess energy.

In which case it will conduct large S-turns to dump lift or will be flown wings-level for maximum range versus drag to extend the flying distance. The Orbiter becomes subsonic (noted by a sonic boom) at 49,000ft, approximately 39 miles from the runway.

Approach and landing phase

The approach and landing phase is encountered at 10,000ft and a speed of 345mph, 7.9 miles from touchdown. With a steep glide slope angle of 17–19°, more than seven times steeper than the approach path for an airliner, auto-land is initiated and in the latter portion of the TAEM the descent rate is about 10,000ft/min, more than 20 times that of a commercial airliner! About 1,750ft above the ground the guidance maintains the Orbiter on the centreline of the runway and positions the vehicle for the pre-flare, where the glide slope angle is reduced to 1.5°, the landing gear being deployed at 100ft.

Final flare occurs at 80ft to a sink rate of 9ft/sec across the runway threshold, waypoint 2 in the auto-land mode, touchdown takes place nominally 2,500ft beyond the runway threshold at a speed of 225–235mph. As the airspeed

bleeds down below 189mph the nose is ever so gently rotated down and the Shuttle has landed, assisted by release of a drag-'chute from the tail of the Orbiter just before the nose wheel engages the runway. It consists of a 9ft diameter pilot 'chute which extracts the 40ft main 'chute trailing 89ft behind the Orbiter, initially in a 40 per cent reefed condition to lessen the shock of retardation. There is one final call of 'wheels stop' from the flight deck before a flurry of activity prepares the Orbiter for shutdown, including shutting down the Auxiliary Power Units about 15min later.

Mission accomplished

There is still much work to be done and a range of vehicles accompany the Orbiter as it rolls to a stop on the 15,000ft runway at the Kennedy Space Center. Hazardous fluids must be drained, liquids removed, inspection carried out to see if the Orbiter has sprung any leaks, and engineers keen to give a cursory glance at the condition of the thermal protection system. Only after it has been 'safed' will the Orbiter be towed to the Orbiter Processing Facility and the cycle will begin again. But for the crew it will have been an unforgettable experience aboard one of the most remarkable flying machines of all time, the world's first reusable Shuttle.

ABOVE If weather prevents tho Orbiter landing at the Kennedy Space Center, it has a secondary landing opportunity at Edwards Air Force Base, California, where it would be mated to the Shuttle Carrier Aircraft for a cross-country flight back to Florida. *(NASA)*

LEFT *Challenger* is carried atop the converted Boeing 747 SCA over the Johnson Space Center, Houston, Texas. *(NASA)*

The missions

A personal reflection by the author

⬛━━⬤━━━━━━━⬛

At the beginning of this book, we reviewed events that changed the way NASA put humans into space. Studies into a Shuttle vehicle formally got under way in 1969, the year two men first walked upon the moon. Before the last of six lunar landings had been accomplished, the prime contractor selected to build the Shuttle was already six months into detailed design studies. Forty years on, it is time to consider events and see how well the Shuttle fulfilled the aspirations of its design team and the hopes of the people that built it.

LEFT Astronaut Thomas Akers conducts an EVA during the STS-49 mission in May 1992, on the first flight of OV-105 *Endeavour*, built to supplement the fleet after the loss of *Challenger* more than six years earlier. *(NASA)*

ABOVE The deployment of satellites and spacecraft was a major activity during the first phase of operations prior to the loss of *Challenger*, the Shuttle's 25th flight. Between 1988 and 1998 that job continued, but only for government satellites such as this TDRS communications relay, which has been of great value in supporting not only Shuttle operations but many unmanned spacecraft also. *(NASA)*

When the Shuttle was designed the world was a very different place. Two giant superpowers strode the geopolitical scene like latter day colossi, each with the intention of changing the world for ever. It was the Cold War, getting more threatening with each passing day. The Vietnam War was still escalating toward its horrifying finality, tensions in the Middle East threatened atomic war, and the Nixon administration was conducting talks with the Soviet Union on limiting the rate at which each country's inventory of nuclear weapons was escalating. The Boeing Jumbo Jet opened a new era in mass air travel, and plans were well advanced for Concorde – the world's first supersonic airliner – to enter service, carrying passengers across the Atlantic at more than twice the speed of sound. The moon race had been won, soon we would be on Mars and Hilton was taking reservations for the first hotel in space.

Against this background the Shuttle was born. It was a stepping stone to routine space transportation ferrying people, cargo, satellites and space station modules to low earth orbit, from where nuclear-powered rocket stages would send the next generation of astronauts to colonise the moon and plunge deeper into the great ocean of space that lay before us.

With unrecognised naivety we all believed the Shuttle to be capable of making up to 60 flights a year – more if needed – and that this winged space plane would be the workhorse of the Space Age hauling everything that went into orbit. In 1979 I published a book about the Shuttle, foretelling the wonderful work it would do as we forged new access routes to space. Captain Chester M. Lee, Director of Space Transportation System Operations, was a man at NASA HQ with whom I worked for several years. He wrote a foreword for that book in which he described the Shuttle as the next great step:

'Space has been called the last "frontier" … excursions to date by men into space have been compared with the Conestoga wagon trips across the American continent in the 19th century… The Space Shuttle is to the space launchers of the past as the train was to the Conestoga wagon of the West. It will be cheaper to operate, be capable of making many trips in its lifetime, and will provide access into a realm previously preserved for a precious few pioneers.'

Those words encapsulate the visionary zeal that permeated the first decade of Shuttle development, but as the 1970s wore on it became apparent that the extravagant expectations of the programme could not be met.

Beyond the dream

Hope that the Shuttle could be turned around within two weeks was impossible to fulfil, and the expectation that it would replace all expendable launch vehicles began to look increasingly difficult to achieve long before the first flight in 1981. The Shuttle was not going to be the cheap route to put payloads into space and technical challenges were proving difficult to overcome. As with any new flight vehicle there were unknown problems waiting to be discovered, and aspects of the design that would lead to difficulties both operationally and in specific flight regimes.

Examples include the selection of thermal insulation tiles that proved difficult to retain in place during launch and re-entry, and were subject to major damage from minor pieces of debris. This was a significant problem during the preparation of the first flight vehicle for launch and did much to contribute to the near two-year delay in getting *Columbia* off the pad. But the pace of development was brisk in a tight financial environment. While management of the Shuttle programme went through significant changes which are not appropriate to delve into here, there were major successes in overcoming technical design and test difficulties.

One of the big success stories of the Shuttle was the SSME – the Orbiter's main cryogenic engines, which were designed with higher operating specifications than had been accepted for any other production engine of their type. With extraordinary demands made upon the theoretical extrapolation of flow dynamics, materials and maintainability of these engines, the challenge could not have been met without great effort in both design and test phases. That was due in no small part to combined input from several companies and in the end two industry giants – Rocketdyne and Aerojet – provided invaluable input to the entire programme. On many occasions almost insuperable problems threatened the pace of development and test, and in the end it all came down to a national effort with key players.

On the record

The Shuttle began as a NASA project in the mid-1960s at the core of a so-called Post-Apollo programme, designed to capitalise on technical, scientific, engineering and management success with events leading up to the moon landings. Inspired by the flight of the world's first spaceman, Russia's Yuri Gagarin, the Apollo moon goal laid down by President John F. Kennedy on 25 May 1961, was an eye-watering challenge and raised the level of space activity by several orders of magnitude. The next goal was how to capitalise on all that development and expansion in capability.

The Post-Apollo plan envisaged a reusable space transportation system lifting everything launched into space, thereby benefiting from an economy of scale in that several dozen flights each year would cut the cost of launching payloads into orbit, freeing up money for development of completely new activities in

space. But even before the Shuttle had been formally approved by the White House, it was apparent that there would be insufficient money to build both the Shuttle and the permanently manned space station it was supposed to support. And as development merged into flight operations, it soon became clear that all hope of a low-cost space transportation system was still a dream and would never be realised by the Shuttle.

The three ages of Shuttle

Because they spanned more than 30 years, Shuttle missions can be divided up in a variety of different ways. To do so by function is perhaps the most logical and by using this method the flight programme can be divided into three very distinct and separate phases: the first 25 missions, in which an attempt was made to increase flight rates at any cost and carry all manner of government and commercial satellites; the next 67 flights during which the Shuttle was restricted to non-commercial payloads, flying military missions and docking with the Russian Mir space station; and the last 43 missions building the International Space Station, at last carrying out the role for which the Shuttle was designed.

BELOW The Shuttle proved itself adept at servicing satellites in space, and at retrieving satellites designed for geostationary orbit but stranded in low earth orbit when their rocket motor failed. Here, on the first mission where three crewmembers worked outside at the same time, astronauts from *Atlantis* attach a new boost motor to the giant Intelsat VI telecommunications satellite on the STS-49 mission in March 1992. (NASA)

Phase 1 – 1981–86

By the late-1970s the role for the Shuttle had changed from space station support vehicle to a launch system for satellites and spacecraft, and for conducting scientific research using equipment in the mid-deck area of the crew compartment, or in a laboratory fixed in the cargo bay. The emergence of a buoyant commercial satellite industry built around communication and TV broadcasting, and the imminent appearance of a new expendable launch vehicle from Europe called *Ariane*, spurred Shuttle managers to attract customers around the world to launch their satellites by Shuttle.

To do that it had to compete with much cheaper, and expendable, rockets so the US government had to massively subsidise Shuttle flights carrying commercial satellites. What had been anticipated as a replacement for expendable vehicles became a financial burden for NASA. Several early flights were only flown to carry these commercial payloads – satellites for Canada, India, Indonesia, Mexico, the Arab league, and Australia in addition to a host of domestic US broadcasters such as AT&T and RCA. But to maintain commercial attractiveness the price these customers paid was a tiny fraction of what it cost NASA to fly the Shuttle.

In its first 25 flights the Shuttle demonstrated what it could have achieved over the long term, carrying satellites into orbit from where strap-on rocket packs sent them into geostationary orbit, retrieving satellites stranded in the wrong orbit by failed boost motors, and rendezvousing with failed satellites so that space-walking astronauts could repair and return them to full operations. In the first five years it seemed that each flight probed further into the corners of its capabilities, promising, albeit at a much slower pace than anticipated, a dramatic new age of reusable space transportation. But all along, the launch of satellites was only a secondary role. Its prime function still lay unrealised.

The first four Shuttle flights carried a crew of only two and were known as OFT (Orbital Flight Test) missions, with the Shuttle carrying arrays of development flight instrumentation to measure the Orbiter's performance. In June 1982 President Ronald Reagan witnessed the landing of *Columbia* after its fourth flight and imprudently declared the Shuttle 'fully operational', which it was not. A flurry of commercial satellite cargoes followed and in November 1983 it carried the first Spacelab scientific laboratory on a flight lasting more than 10 days.

Spacelab had been financed by the European Space Agency (ESA), with 52 per cent paid for by Germany. It comprised a pressurised module built in Germany, which could be carried inside the Shuttle Orbiter, and a set of external pallets for carrying scientific equipment built in the UK attached to the rear of the module inside the cargo bay. Spacelab resulted from discussions with European countries in 1970 and a firm agreement was reached in 1973, a year after NASA got formal approval to build the Shuttle.

The idea had been that ESA would pay for design, development and fabrication of one Spacelab pressure module and associated pallets (about US$1 billion) and that the US would buy four more pressure modules at US$250 million each. Net result – Europe would gain expertise with manned space flight and get its money back while gaining the chance to fly European astronauts in the Shuttle. NASA would get to do science in a laboratory module that its own budget was incapable of buying, deferring the cost of expanding the

BELOW The International Space Station has become the iconic symbol of cooperation in space, transforming excursions beyond earth from an ideological race into a new frontier for all humans to explore together. The ISS could not have been built, or stocked with spares, without the Shuttle. *(NASA)*

number of Spacelab modules available until the development cost of Shuttle had been paid for. Then NASA could go about planning for a permanently manned space station left in earth orbit, routinely serviced and replenished by the Shuttle with crew and equipment.

But developing the Shuttle took longer than planned. When NASA went to industry in 1970 for Phase B definition proposals it wanted the Shuttle flying into space operationally by 1977. After North American won the contract in 1972, developing the Shuttle proved harder than anticipated and an ever widening net of national resources was embraced by NASA and the manufacturers to solve problems and test new technology. By 1974 more than 30,000 people were working on the Shuttle; by 1977 more than 50,000. But the first flight had slipped into late 1979 and still the delays kept piling up.

Expectations that the Shuttle could be turned around and returned to space within two weeks, that each Orbiter could make 25 flights a year into space and that overall the Orbiters would be flying in excess of 65 missions a year soon proved wildly unrealistic. After the first flight in 1981 a different mood set in, but even then annual flight rates of 48 Shuttle missions a year were being predicted, around half of them carrying Spacelab modules preparing a new generation of scientists to work in the permanently manned space station that NASA pinned its future hopes on.

After the first Spacelab mission in November 1983 on the ninth Shuttle flight, there were three more in 1985, by which time, in January 1984, President Reagan had challenged NASA to build a permanently manned space station 'and to do it within a decade'. This was the station it had been impossible to develop in parallel with the Shuttle and that would vindicate development of the space plane. While that was good news for NASA, it brought a mixed message to the Europeans. Now there would be no need – or money – to buy four more Spacelab modules, making the first one virtually a gift to the USA, while NASA money went on the space station. Reagan named it *Freedom* and sent NASA boss James Beggs on a globetrotting tour of non-communist countries asking for their financial and material support in building it.

Eventually, the Europeans agreed to participate. Before that, on 28 January 1986, *Challenger* exploded within sight of thousands watching from Cape Canaveral and brought the first phase of Shuttle operations to an abrupt end.

Phase 2 – 1988–98

It took 32 months to get the Shuttle back into space and in that time a lot had changed. The Presidential Commission set up to investigate the disaster demonstrated that it was no accident but the consequences of a dangerous game of Russian roulette – to keep flying on-time, every-time, as a very senior NASA administrator told this author only a few months prior to the loss of *Challenger*. The findings all but destroyed the national faith held in NASA by a grateful public, shocked by revelations that smacked of gambling with lives for faster flight rates and satisfied satellite customers.

By the time flights resumed in September 1988 all commercial satellite traffic had been banned from flying in the Shuttle. The commercial pressures were just too great to push the Shuttle to match the pace of less complicated expendable rockets. Europe's *Ariane* was proving a great success at putting satellites into orbit and the Air Force turned its back on the Shuttle and reopened the

ABOVE The crew aboard *Discovery* cavorts inside the Spacelab module secured inside the payload bay. Built in Europe, Spacelab provided the experience from which the European Space Agency could participate in the International Space Station, with the *Columbus* laboratory module that will continue to provide research facilities for more than a decade to come. *(NASA)*

production lines for expendable rockets, subsidising a commercial launch vehicle industry that thrives today. The Air Force had some payloads that could only fly on the Shuttle and lifted those on eight dedicated military flights between December 1988 and December 1992. Added to the three launched before *Challenger*, a total of just 11 Air Force missions were flown by the Shuttle, which it had once claimed would be its workhorse for space.

The timely withdrawal of US Air Force use of the Shuttle sounded the death knell for a second launch site, one located at Vandenberg Air Force Base (VAFB), California, situated on the Pacific coast north of Los Angeles. Since 1972 VAFB had been assigned to launching all Shuttle missions into orbits with an inclination greater than 57°, to prevent the ascending Shuttle flying over the populated eastern seaboard of the continental United States. Heading south into high inclination and polar orbits the Shuttle would fly over water and not land. While NASA, too, would have sent some missions from VAFB, the Air Force was driving the requirement because it needed to launch and service certain classified missions at these high inclinations. The Air Force would have been the primary customer. The *Challenger* disaster changed those plans and the Air Force abandoned its use of the Shuttle. *Enterprise* had been delivered to VAFB for a fit-check on the assigned launch complex (SLC-6, pronounced 'slick six') but the idea of launching payloads from VAFB had been abandoned by the end of 1986.

A consequence of the loss of *Challenger* was the manufacture of a sixth Orbiter as replacement. Congress approved supplementary funding and in July 1987 NASA commissioned Rockwell International to build OV-105 *Endeavour*, from a set of spare parts comprising major vehicle sections held in reserve should they be needed to mend seriously damaged Orbiters. Final assembly was completed in July 1990 with rollout nearly 10 months later. Its first flight was on 7 May 1992, three years and eight months after Shuttle flights had resumed.

In this second phase of Shuttle operations only government missions were flown, scientific satellites, unmanned astronomical observatories and the awe-inspiring Hubble space telescope, lifted into orbit on 24 April 1990. Giant spacecraft were launched to Venus, to Jupiter and to Saturn and a wide variety of technological and scientific experiments were carried out. The Shuttle was safer now, but still plagued with recurring incidents and technical problems sufficiently hidden to beguile the unwary into believing it could perhaps become a fully operational transportation system. It never was and the post-*Challenger* reality stripped it of much of the pressure that had underpinned the disaster.

This phase of Shuttle missions prepared the way for its definitive role – that of lifting into orbit all the many elements of a permanent space station called *Freedom*. Toward that end, more Spacelab missions were conducted. There had been four before *Challenger* and 17 more between December 1990 and April 1998. Two were dedicated missions packed with German experiments and interspersed were five Spacehab flights, a commercial venture leasing rack space for science experiments in what looked like a min-Spacelab attached to the forward end of the cargo bay for direct access by astronauts. But the real jewel in the list of achievements began in 1995, which would transform space station *Freedom* and unite former adversaries in the race for space.

After the collapse of the Berlin Wall and the subsequent implosion of the Soviet regime in Russia, President Bill Clinton made it a purposeful aim of the State Department to extend a hand of cooperation to the former Soviet Union. Since Reagan announced his goal of a space station in 1984, every year NASA had been short-changed by Congress on the money it needed to press ahead. For 10 years *Freedom* limped along, studied endlessly, but with no launch date in sight for the first elements of a structure that would take several years to assemble. Inviting its former enemy to become a principal partner in the space station, the US brought Russia in to a great international venture that now included Europe, Japan and Canada.

In preparation for working closely on what would henceforth be renamed the International Space Station, or ISS, NASA ran a rendezvous mission to the Russian Mir space station in

February 1995. Then, between June 1995 and June 1998, it carried out eight docking flights to Mir where US astronauts and Russian cosmonauts worked together on scientific experiments in space. What began as the next stage on a race with the Soviet Union in the late 1960s had outlived the end of the Cold War, and forged an inseparable union with a former adversary. The second phase of Shuttle missions ended with those events.

Phase 3 – 1998–2011

In more than 40 flights beginning with *Endeavour* on 4 December 1998, the Shuttle has systematically launched the elements of what has been hailed as the greatest engineering venture of all time, arguably the greatest test of human cooperation in the most inhospitable environment encountered by humans. It is this for which the Shuttle was designed and built and this will be its greatest legacy. In this period, a second Orbiter was destroyed when on 1 February 2003 *Columbia* broke apart on re-entry due to damage to its thermal insulation. And on other flights the Hubble telescope has been routinely serviced, upgraded, improved and given a new lease of life. Undoubtedly, the loss of *Columbia* stimulated reappraisal of the Shuttle. This time there would be no replacement Orbiter and a decision was made shortly thereafter to retire it after completion of its last and most important role, the assembly of the International Space Station (ISS).

It is the ISS that has justified the Shuttle. The first element, called *Zarya,* was launched by the Russians on 20 November 1998, by Proton rocket, followed on 4 December with the launch of *Endeavour* carrying the *Unity* module, the first of many that would make up the ISS. The first US pressurised laboratory module, called *Destiny,* was lifted by *Atlantis* on 7 February 2001. Building on the experience with Spacelab, Europe's *Columbus* module was lifted to the ISS on 2 February 2007, followed by Japan's module on 31 May 2008.

With a length of 167ft, a width of 357ft, a height of 66ft and a weight of more than 400 tons, it orbits the earth at 17,500mph and provides habitable space for a permanent crew of six from countries around the world.

In the beginning, the Shuttle researched the possibility of producing new vaccines in the weightlessness of space, something that was impossible to manufacture in the gravity at earth's surface. New semiconductor crystals were grown five times bigger than anything that could grow on earth. And in conducting medical experiments in space, where astronaut's bodies are overwhelmed by the same processes of deterioration that afflict senior citizens on earth, new methods on how to keep people healthy for longer is here now and being applied in hospitals around the world.

When evaluating the benefits to humans around the planet, it is difficult to count the cost of what has been made possible through the Shuttle and the ISS it has helped build. From forging working relationships with former adversaries to unlocking secrets about the human body and the way it can be healed, this work goes on day after day by scientists and research workers flying to the space station. The operational history of the Shuttle is now at an end, its work done. It leaves a legacy and a wealth of engineering knowledge that will outlast the generation that built it. Of all the things with wings – this truly is the most magnificent of all flying machines.

BELOW Prophetically a portent of the end of Shuttle operations, this sign was held up by Dale Gardner during a space walk on the STS-51A flight, the second for *Discovery*, in November 1984, after the retrieval of the Palapa B and Westar VI communications satellites that had become stranded in earth orbit when their boost motors failed. *(NASA)*

Appendix 1

SPACE SHUTTLE FLIGHTS BY ORBITER

The following missions are in chronological order by Orbiter vehicle and mission numbers are in the sequence designated and not in the order flown.

COLUMBIA (OV-102)

Mission	Launch Date	Mission Information
STS-1	04.12.81	1st Shuttle Orbital Flight Test (OFT-1)
STS-2	11.12.81	2nd Shuttle Mission/Office of Space and Terrestrial Applications-1 (OSTA-1)
STS-3	03.22.82	3rd Shuttle Mission/Office of Space Science-1 (OSS-1)
STS-4	06.27.82	Department of Defense, materials experiments
STS-5	11.11.82	First commercial comsats (ANIK C-3 and SBS-C) by Shuttle
STS-9	11.28.83	Orbital Laboratory and Observations Platform/First Spacelab mission
STS-61C	01.12.86	SATCOM KU-1 comsat
STS-28	08.08.89	Two Department of Defense comsats (DSCS-III F-2 and F-3)
STS-32	01.09.90	SYNCOM IV-F5 comsat; LDEF retrieval (see Challenger STS-41C)
STS-35	12.02.90	Spacelab ASTRO-1 astronomy mission
STS-40	06.05.91	Spacelab Life Sciences-1
STS-50	06.25.92	Spacelab microgravity research USML-1
STS-52	10.22.92	USMP-1; LAGEOS II
STS-55	04.26.93	Spacelab D-2 German mission
STS-58	10.18.93	Spacelab SLS-2 life sciences
STS-62	03.04.94	USMP-2, OAST-2 science mission
STS-65	07.08.94	Spacelab IML-2 international microgravity mission
STS-73	10.20.95	Spacelab microgravity research USML-2
STS-75	02.22.96	Tethered satellite (TSS-1R), US Microgravity Payload-3 science mission
STS-78	06.20.96	Spacelab life and microgravity (LMS) science
STS-80	11.19.96	Free-flying science satellites (ORFEUS-SPAS II and WSF-3)
STS-83	04.08.97	Spacelab MSL-1 material science mission
STS-94	07.01.97	Spacelab MSL-1 material science re-flight mission
STS-87	11.19.97	US Materials Processing-4 mission; Spartan 201-04 free-flying satellite
STS-90	04.17.98	Final Spacelab mission, neurological sciences mission
STS-93	07.23.99	Chandra X-Ray Observatory
STS-109	03.01.02	Hubble Space Telescope servicing mission
STS-107 *	01.16.03	Microgravity research mission/SPACEHAB

* Columbia and crew were lost on 02.01.03 during re-entry over East Texas approximately 16 minutes before landing.

CHALLENGER (OV-99)

Mission	Launch Date	Mission Information
STS-6	04.04.83	Tracking and data relay satellite-1 (TDRS-1)/ 1st Shuttle space walk
STS-7	06.18.83	Indonesian and Canadian comsats/1st US woman in space
STS-8	08.30.83	Multipurpose satellite launch for India/1st night launch and landing
STS-41B	02.03.84	US and Indonesian comsats, Manned Manoeuvring Unit, 1st KSC landing
STS-41C	04.06.84	Long Duration Exposure Facility deploy (LDEF), 1st on-orbit spacecraft repair
STS-41G	10.05.84	US Earth Radiation Budget Satellite (ERBS)
STS-51B	04.29.85	Spacelab-3 microgravity and life sciences mission
STS-51F	07.29.85	Spacelab-2 solar physics mission
STS-61A	10.30.85	Spacelab D-1 (1st German dedicated Spacelab)
STS-51L**	01.28.86	TDRS-2 satellite, SPARTAN-203 free-flyer

**Challenger and crew were lost on 01.28.86 approximately 73 seconds after lift-off, following a catastrophic explosion in the right solid rocket booster.

BELOW Bruce McCandless flies the Manned Manoeuvring Unit backpack while un-tethered from *Challenger* during the STS-41B mission in February 1984. *(NASA)*

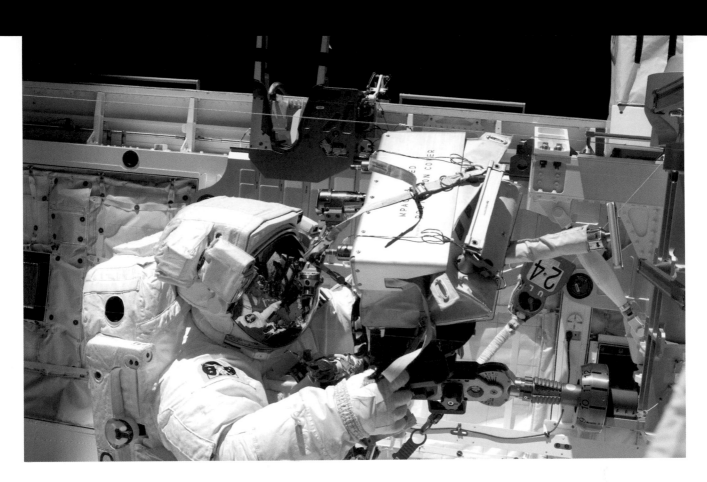

DISCOVERY (OV-103)		
Mission	Launch Date	Mission Information
STS-41D	08.30.84	SBS-D; TELSTAR 3C, SYNCOM IV-2 comsats; Solar Wing experiment
STS-51A	11.08.84	Canadian comsat Anik D2, Leasat-1 US Navy comsat
STS-51C	01.24.85	Orion 1 signals intelligence satellite for Defense Department
STS-51D	04.12.85	Canadian comsat Anik C3, Leasat-3 US Navy comsat
STS-51G	06.17.85	MORELOS-A, ARABSAT-A and TELSTAR-3D comsats
STS-51I	08.27.85	US ASC-1 satellite, Australia's AUSSAT-1, Leasat-2 US Navy comsat
STS-26	09.29.88	TDRS-C launch
STS-29	03.13.89	TDRS-D launch
STS-33	11.22.89	Orion 2 signals intelligence satellite for Defense Department
STS-31	04.24.90	Hubble Space Telescope deploy
STS-41	10.06.90	Ulysses spacecraft to Jupiter, several science experiments
STS-39	04.28.91	Department of Defense experimental research mission
STS-48	09.12.91	Upper Atmosphere Research Satellite
STS-42	01.22.92	Spacelab IML-01 microgravity mission
STS-53	12.02.92	SDS B3 comsat on last dedicated Department of Defense mission
STS-56	04.08.93	ATLAS-2 earth atmosphere science mission, free-flying SPARTAN-201
STS-51	09.12.93	Comsat technology satellite, astronomy science (ORFEUS-SPAS)
STS-60	02.03.94	Free-flying physics experiment (WSF-1) SPACEHAB-2
STS-64	09.09.94	Physics experiment LITE, and free-flying SPARTAN-201

ABOVE Launched aboard *Discovery* in April 2010, Clayton Anderson works on equipment during assembly of the International Space Station. *(NASA)*

DISCOVERY (OV-103) - continued		
Mission	Launch Date	Mission Information
STS-63	02.03.95	SPACEHAB-3 materials and life sciences mission
STS-70	07.13.95	TDRS-G
STS-82	02.11.97	2nd Hubble Space Telescope servicing
STS-85	08.07.97	Earth science mission (CRISTA-SPAS-02)
STS-91	06.02.98	9th and final Shuttle-Mir docking
STS-95	10.29.98	John Glenn's flight; SPACEHAB
STS-96	05.27.99	2nd International Space Station flight
STS-103	12.19.99	3rd Hubble Space Telescope servicing mission
STS-92	10.11.00	International Space Station assembly flight 3.3A
STS-102	03.08.01	International Space Station assembly flight 5A.1
STS-105	08.10.01	International Space Station assembly flight 7A.1
STS-114	07.26.05	International Space Station assembly flight LF1
STS-121	07.04.06	International Space Station assembly flight ULF1.1
STS-116	12.09.06	International Space Station assembly flight 12A.1
STS-120	10.23.07	International Space Station flight 10A
STS-124	05.31.08	International Space Station flight 1J
STS-119	03.15.09	International Space Station flight 15A
STS-128	08.28.09	International Space Station flight 17A
STS-131	04.05.10	International Space Station flight 19A
STS-133	02.24.11	International Space Station ULF

ATLANTIS (OV-104)

Mission	Launch Date	Mission Information
STS-51J	10.03.85	Department of Defense communications satellite mission
STS-61B	11.26.85	Mexican (MORELOS-B), Australian (AUSSAT-2), US SATCOM KU-2
STS-27	12.02.88	Lacrosse radar imaging satellite for Department of Defense
STS-30	05.04.89	*Magellan* spacecraft to Venus
STS-34	10.18.89	*Galileo*; Shuttle Solar Backscatter Ultraviolet Experiment
STS-36	02.28.90	Mysty reconnaissance satellite for Department of Defense
STS-38	11.15.90	SDS-B2 comsat for Department of Defense
STS-37	04.05.91	Gamma Ray Observatory
STS-43	08.02.91	TDRS-E, physics experiments
STS-44	11.24.91	Early warning satellite (DSP-F-16) for Department of Defense
STS-45	03.24.92	Atmospheric sciences mission (ATLAS-1)
STS-46	07.31.92	Tethered satellite (TSS-1), EURECA free-flyer deployed
STS-66	11.03.94	Atmospheric sciences mission (ATLAS-3), CRISTA-SPAS
STS-71	06.27.95	1st Shuttle-Mir docking
STS-74	11.12.95	2nd Shuttle-Mir docking
STS-76	03.22.96	3rd Shuttle-Mir docking; SPACEHAB
STS-79	09.16.96	4th Shuttle-Mir docking
STS-81	01.12.97	5th Shuttle-Mir docking
STS-84	05.15.97	6th Shuttle-Mir docking
STS-86	09.25.97	7th Shuttle-Mir docking
STS-101	05.19.00	3rd International Space Station flight
STS-106	09.08.00	International Space Station Flight 2A.2b
STS-98	02.07.01	International Space Station assembly flight 5A
STS-104	07.12.01	International Space Station assembly flight 7A
STS-110	04.08.02	International Space Station 8A
STS-112	10.07.02	International Space Station 9A
STS-115	09.09.06	International Space Station assembly flight 12A
STS-117	06.08.07	International Space Station
STS-122	02.07.08	International Space Station/ESA *Columbus* laboratory
STS-125	05.11.09	Hubble Space Telescope final servicing
STS-129	11.16.09	International Space Station
STS-132	05.14.10	International Space Station ULF
STS-135	07.08.11	International Space Station MPLM

ENDEAVOUR (OV-105)

Mission	Launch Date	Mission Information
STS-49	05.07.92	Intelsat VI repair
STS-47	09.12.92	Spacelab-J dedicated Japanese experiments
STS-54	01.13.93	TDRS-F, X-ray spectrometer
STS-57	06.21.93	SPACEHAB-1; EURECA retrieval
STS-61	12.02.93	1st Hubble Space Telescope servicing mission
STS-59	04.09.94	Space Radar Laboratory (SRL-1)
STS-68	09.30.94	Space Radar Laboratory (SRL-2)
STS-67	03.02.95	Ultraviolet Observatory science mission (ASTRO-2)
STS-69	09.07.95	Free-flyer SPARTAN 201-03, physics experiment (WSF-2)
STS-72	01.11.96	Science and technology mission (SFU), OAST-Flyer
STS-77	05.19.96	SPACEHAB; SPARTAN (IAE)
STS-89	01.22.98	8th Shuttle-Mir docking
STS-88	12.04.98	1st International Space Station flight
STS-99	02.11.00	Shuttle Radar Topography Mission
STS-97	11.30.00	International Space Station assembly flight 4A
STS-100	04.19.01	International Space Station assembly flight 6A
STS-108	12.05.01	International Space Station assembly flight UF-1
STS-111	06.05.02	International Space Station UF2
STS-113	11.23.02	International Space Station 11A
STS-118	08.08.07	SPACEHAB; ISS CMG replacement
STS-123	03.11.08	Japanese Kibo module & Canadian manipulator (Dextre) to ISS
STS-126	11.14.08	International Space Station upgrade
STS-127	07.15.09	International Space Station Japanese Kibo assembly
STS-130	02.08.10	International Space Station assembly
STS-134	05.16.11	International Space Station ULF

ENTERPRISE (OV-101)

Enterprise was used for Air-Launch Flight Tests in 1977 from the top of the converted Boeing 747. It is on display at Dulles Airport outside Washington, DC.

BELOW Mission Control during the flight of STS-41G, in October 1984, during which *Challenger* conducted a variety of scientific experiments. *(NASA)*

Appendix 2

GLOSSARY AND ABBREVIATIONS

ALT Air Launched Tests, five drop-tests conducted by Enterprise in 1977

AM Airlock Module, located either in aft end of mid-deck or forward end of payload bay

AOA Abort Once Around, late ascent abort going round the earth to land

APU Auxiliary Power Unit, provides power for the hydraulic units

Avionics An amalgam of 'aviation' and 'electronics'

Beanie Cap Used to channel oxygen boil-off from top of the ET

BTU British Thermal Unit, energy needed to raise 1lb of water 1°F, 1 BTU=1,055.05585 joules

CCTV Closed Circuit Tele-Vision, cameras plus relay and/or recording system

CM Command Module, Apollo pressurised re-entry vehicle

CO$_2$ Carbon dioxide, two oxygen atoms covalent with one carbon atom

Columbus ESA module attached to International Space Station

Communications blackout
A plasma sheath enveloping the Orbiter temporarily blocking radio frequencies

CRT Cathode Ray Tube, vacuum tube for display monitor

Cryogenic Super-cold fluids below -238°F

CT Crawler Transporter, used to move MLP and Shuttle to the launch pad

CWS Caution & Warning System, alarm sensors and warning devices

Digital Autopilot
Digital control system for automated functions

DOP Diver-Operated Plug, used to plug the aft nozzle on SRBs after splashdown

E-I Entry Interface, the point at which atmospheric deceleration reaches 0.05g

ECLSS Environmental Control & Life Support System, maintaining and reconstituting a living environment

EMU Extra-vehicular Mobility Unit, space suit

ESA European Space Agency, group of European countries formed in 1975, which now includes 18 countries

ET External Tank, incorporating separate tanks for SSME propellants

EVA Extra-Vehicular Activity, space-walking

Flight Control System
Automated commanded flight sequence

FDF Flight Data File, reference manuals for systems and operations

FRF Flight Readiness Firing, a brief test-firing of SSMEs on the launch pad for new Orbiter vehicles

FRCS Forward Reaction Control System, cluster of 16 thrusters in forward fuselage

FRSI Felt Reusable Surface Insulation, thermal insulation blanket

FSS Fixed Service Structure, attached to LC-39 supporting service arms and RSS

Fuel Cell An electricity production system based on reverse electrolysis

GNC Guidance, Navigation & Control, attitude and position guidance

GPC General Purpose Computer, five of which are used In the Orbiter

HAC Heading Alignment Cylinder, imaginary cylinder in the sky around which radius the Orbiter flies to a landing point

High Bay Area in the VAB where full Shuttle stack is erected

HPFTP High Pressure Fuel Turbo Pump, for hydrogen fuel in SSME

HPOTP High Pressure Oxidiser Turbo Pump, for oxygen in SSME

HRSI High-temperature Reusable Surface Insulation, thermal protective tile

HUD Head-Up Display, transparent projection of selected information in the pilot's forward line of sight

ILRV Integral Launch & Re-entry Vehicle, late-1960s concept for shuttlecraft

ISS International Space Station, built and assembled by NASA, ESA, Russia, Japan and Canada

JSC Johnson Space Center, renamed from MSC in honour of President Lyndon B. Johnson

Kbps Kilo-Bits Per Second

KSC Kennedy Space Center, a portion of leased land at Cape Canaveral for NASA and commercial launches named in honour of President John F. Kennedy

Ku-Band Radio frequencies in the 11–18GHz band

LC Launch Complex, designation system for launch pad area with an appended number consisting of one or more launch pads

LCC Launch Control Center, four firing rooms for launch control

LCD Liquid Crystal Display, flat electronic display system superseding CRT

LCG Liquid Cooled Garment, worn under space suit for body temperature control

LH$_2$ — Liquid Hydrogen, fuel for combustion or as reactant in fuel cells

Low Bay — Area in the VAB from where Shuttle elements are brought to the High Bay

LOX — Liquid Oxygen, used as oxidiser for combustion, as reactant in fuel cell or for crew compartment pressurisation

LPFTP — Low Pressure Fuel Turbo Pump, for hydrogen fuel in SSME

LPOTP — Low Press Oxidiser Turbo Pump, for oxygen in SSME

LRSI — Low-temperature Reusable Surface Insulation, thermal protective tile

LWT — Light Weight Tank, reduced weight ET

MCC — Mission Control Center, located within the JSC

MCDS — Multifunction CRT Display System, full suite of electronic equipment for displays on flight deck

MEC — Main Engine Controller, for SSME

MEDS — Multifunction Electronic Display System, glass cockpit technology

MSC — Manned Spacecraft Center, NASA field centre for manned space flight, renamed JSC, close to Houston, Texas

MLP — Mobile Launch Platform, on which the Shuttle is stacked

MPLM — Multi-Purpose Logicstics Module

MVGVT — Mated Vertical Ground Vibration Test, facility at MSC for dynamic testing on mated Shuttle elements

NACA — National Advisory Committee for Aeronautics, formed in 1915 for US Government aeronautical research

NASA — National Aeronautics & Space Administration, successor to NACA from 1 October 1958

O&C — Building Operations & Checkout Building, crew quarters

OBSS — Obiter Boom Sensor System, 50ft rigid, twin-section, boom for viewing exterior of Shuttle

OFT — Orbital Flight Test, first four Shuttle missions in which two active ejection seats were carried by the Orbiter

OMRF — Orbiter Modification & Refurbishment Facility, where Orbiters undergo major maintenance

OMS — Orbital Manoeuvring System, two rocket motors for orbit changes

OPF — Orbiter Processing Facility, at KSC northwest of VAB

OPS — Operational Programme Sequencer, discrete set of software instructions for a specific flight phase

Orbiter — Winged shuttlecraft

OV — Orbiter Vehicle, production allocation preceding a number in order of manufacture

P&W — Pratt & Whitney, engine manufacturer

PASS — Primary Avionics Software System, advanced software package at the core of the GNC

PC — Payload Canister, used to deliver payloads to the Shuttle on the pad via the RSS

PCR — Payload Change-out Room, incorporated within the RSS to receive the PC

RAM — Random Access Memory, computer memory that can be accessed in a random fashion on demand

RCC — Reinforced Carbon-Carbon, a high-temperature thermal insulation

RCS — Reaction Control System, comprising 44 rocket thrusters for attitude control and rendezvous manoeuvres

RCSM — Reaction Control System Module, containing groups of RCS thrusters

RMS — Remote Manipulator System, articulated manipulator arm

RSS — Rotating Service Structure, rotates through 120° to enclose Shuttle on the pad

RTLS — Return To Launch Site, abort mode during early phase of ascent

S-Band — Radio frequencies in the 2–4GHz band

SCA — Shuttle Carrier Aircraft, consisting of two converted Boeing 747 aircraft for carrying the Shuttle

SFDSS — Smoke Detection & Fire Suppression System, smoke detection and fire extinguishers

SOFI — Spray-On Foam Insulation, used as an external thermal insulator on the ET

Spacelab — Module built by ESA for on-orbit scientific research

Squib — Miniature explosive device

SRB — Solid Rocket Booster, comprising SRM and all associated elements for launch

SRM — Solid Rocket Motor, propulsion elements within SRB

SSME — Space Shuttle Main Engine, the three liquid propellant engines at the rear of the Orbiter

STA — Structural Test Article, STA-099 for strength tests converted to OV-099 Challenger

STDN — Spaceflight Tracking & Data Network, ground based tracking and communications system

STS — Shuttle Transportation System, generic name for Shuttle programme and designated abbreviation for Shuttle flights (STS-1, STS-2, etc)

SLWT — Super Light Weight Tank, lightest ET design flown

TAEM — Terminal Area Energy Management, balancing lift and drag during the final stages of descent

TAL — Transoceanic Abort Landing, abort during ascent to a landing in Africa or Europe

TDRS — Tracking & Data Relay Satellite System, geostationary satellites for communications relay

ULF — Utility and Logistics Flight

UTC — United Technologies Corporation

VAB — Vehicle Assembly Building, at KSC for stacking launch vehicle stage and spacecraft

VAFB — Vandenberg Air Force Base, located on the Californian coast

VCPF — Vertical Cargo Processing Facility, area at KSC where payloads are installed in the PC

WCS — Waste Collection System, the Shuttle waste water collection unit

Appendix 3

USEFUL CONTACTS

USA

Kennedy Space Center Visitor Complex
SR405 Kennedy Space Center
Florida 32899, USA
Tel +1 866 737 5235
www.kennedyspacecenter.com
Situated close to the Kennedy Space Center from where NASA launched all its 166 manned space missions, this is a large parkland of exhibits, displays, rockets, Shuttle mock-up to enter, exhibits, displays and guided tours. IMAX theatre and every day outdoor informal lunch with an astronaut.

National Air & Space Museum
6th & Independence Avenue SW
Washington DC 20560, USA
Tel +1 202 633 1000
www.nasm.si.edu
The world's largest aerospace museum with numerous galleries and exhibits covering the space programme and the Shuttle specifically.

Space Center Houston
1601 NASA Parkway
Houston, Texas 77058, USA
Tel +1 281 244 2100
www.spacecenter.org
Contains one of the world's largest collections of space exhibits with tours of Shuttle related facilities right alongside the NASA Johnson Space Center.

Steven F. Udvar Haizy Center
14390 Air & Space Museum Parkway
Chantilly, VA 20151, USA
Tel 001 202 633 1000
www.nasm.si.edu/udvarhaizy
Aerospace museum with the Shuttle Orbiter OV-101 Enterprise as the lead display.

US Space and Rocket Center
One Tranquility Base
Huntsville, Alabama 35805, USA
www.spacecamp.com/museum
Contains one of the largest displays of Shuttle-related hardware and exhibits anywhere and provides educational opportunities with IMAX theatre.

UNITED KINGDOM

British Interplanetary Society
27/29 South Lambeth Road
London SW8 1SZ
UK
Tel 020 7735 3160
www.bis-spaceflight.com
Signatory to the International Astronautical Federation, the BIS was formed in 1933 with Arthur C. Clarke as an early member. The Society is open to all and has two publications available on subscription. It holds regular meetings with lectures and has a library open to members.

National Space Centre
Exploration Drive
Leicester LE4 5NS
UK
Tel 0845 605 2001
www.spacecenter.co.uk
Provides an all-round educational experience in many separate aspects of space research and exploration.

The Science Museum
Exhibition Rd
South Kensington
London SW7 2DD
UK
Tel 0870 870 4868
www.sciencemuseum.org.uk
Contains a full space gallery with many relevant exhibits including a Shuttle model and artifacts from the space programmes of Europe, the UK, the US, Russia and China.

FRANCE

International Space University
Parc d'Innovation
1 rue Jean-Dominique Cassini
67400 Illkirch-Graffenstaden
France
Tel +33 3 88 65 54 30
www.isunet.edu
Provides graduate-level educational resources, courses and qualifications for the aspiring space professional, having graduated more than 3,000 students from 100 countries.

Index